Laser Processing of Materials in Japan:

Research, Development and Applications

Japan Technical Information Service

Elsevier Advanced Technology
Mayfield House, 256 Banbury Road, Oxford OX2 7DH, UK

British Library Cataloguing in Publication Data
Laser processing of materials in Japan.
1. Materials. Processing. Use of lasers
I. Japan Technical Information Service
620.112

ISBN 1-85617-036-5

Contents

Introduction 1

1. **Government Scientific and Technological Policies for Lasers** 3

1.1 National projects 3

1.2 Subsidies 5

1.3 Joint study by universities and private enterprises 5

2. **Activities by Societies** 8

3. **Governmental Scientific and Technological Policies** 10

3.1 Large-scale resarch and development projects by the MITI's Agency of Industrial Science and Technology 10

3.2 Research and development using the science and technology promotion adjustment working fund 30

3.2.1 Research goal 30

3.2.2 Outline of research 34

3.3 Special research funding 36

3.3.1 Research on control technology of the laser chemical reaction by the Chemical and Technology Laboratory 36

3.3.2 Research on new laser technology by the Institute of Physical and Chemical Research 38

3.4 Laser research by the Institute of Physical and Chemical Research 39

3.4.1 Members of the Laser Science Research Group 39

3.4.2 Reports 43

3.5 Summary 50

4. **Technological Details** 50

4.1 Present status and future trends in laser processing technology 51

4.1.1 Laser cutting 58

4.1.2 Laser welding 63

4.1.3 Experimental laser processing 66

4.1.4 New experiments 69

4.2 Creation of new materials and substances 72

4.2.1 Film deposition by the light excitation process 73

4.2.2 Laser annealing (high-speed crystal growth) 79

4.2.3 Photo-etching 80

4.2.4 Materials extraction, refining and synthesis 81

4.2.5 Metal alloys . . . 84
4.2.6 Summary . . . 85
4.3 Laser generator . . . 88

5. **Markets Trends** . . . 91
5.1 Prospects for the machine tool industry . . . 91
5.1.1 Demand trends . . . 91
5.1.2 Technological trends . . . 94
5.2 Trend in the laser processing machine market . . . 94

6. **Conclusion** . . . 99
References . . . 102

Introduction

In 1966, laser technology was first applied in the United States to perforate diamond dies. In Japan, the Laser Process Research Society was set up in 1972, the Laser Technology Social Gathering was established in 1973, and a national project, "Flexible Manufacturing System Complex provided with Laser ", was created by the Ministry of International Trade and Industry (MITI) and the Agency of Industrial Science and Technology in 1978. These Japanese entities have contributed to the rapid development of laser technology in the last few years.

Nowadays, markets demand flexible, small-lot production using new materials that often possess minimal workability. This necessitates the energetic development of a laser technology for the establishment of a laser-implemented Flexible Manufacturing System (FMS).

In Japan, the development of advanced technologies is often promoted by the government. Laser development is no exception.

Figure 0.1 shows changes in the number of laser-related (IPC/B23K26) patents applied for in Japan since 1975.

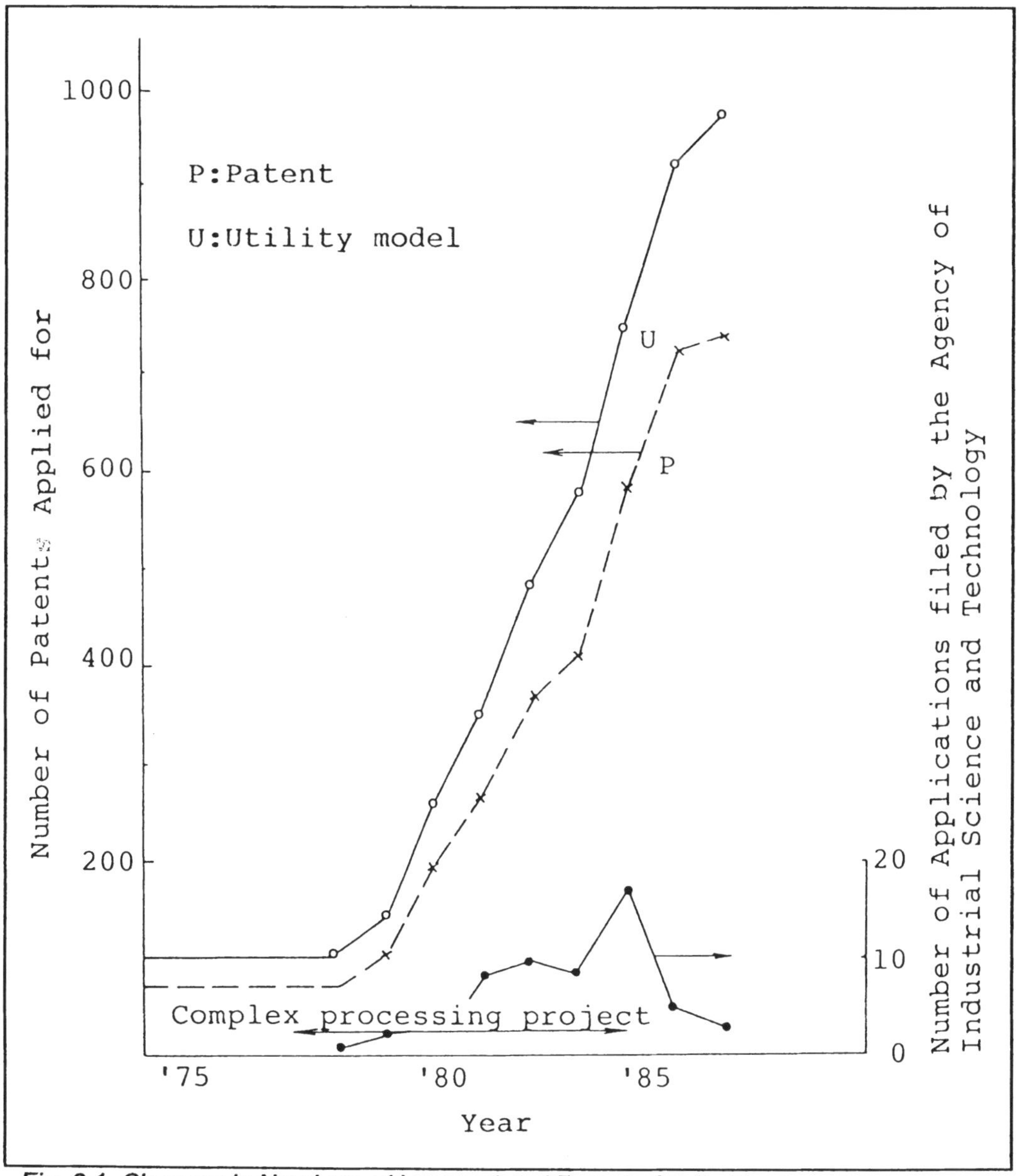

Fig. 0.1 Changes in Numbers of Laser-related Patents Applied for

The Figure shows that the number of patents applied for, which had been fewer than 100 per year, rapidly increased since the start of the project. But the number of patents applied for by the project set up by the Agency of Industrial Science and Technology is less than 20, even in the last year of the project. This illustrates that the project was an indirect trigger to the development of Japanese laser technology.

The present report focuses on the status of Japanese laser technology development in connection with R&D promotion policies, along with the present and future market status.

1. Government Scientific and Technological Policies for Lasers

1.1 National projects

As part of the government's scientific and technological promotion policies, research is being done on the improvement of laser performance, including enhanced output power and higher efficiency, and the development of new lasers employing ultraviolet rays. In addition, R&D activities are in progress on: laser-energy utilization; applications for laser-chemical reactions; and applications for analysis, instrumentation, and control.

A national project, the "Flexible Manufacturing System Complex provided with Laser", was started by MITI in 1978 and ended in 1985. The Electrotechnical Laboratory, however, which played a central role during that period, has continued its research. The Laboratory gained a worldwide reputation by introducing "Flexible Manufacturing with Lasers" to an international symposium held at Boston in 1984.

As a research project with three more year period, nongovernmental organizations have been renting the research facilities for the said national project at Tsukuba Laboratory. Management fees are paid by private organizations. These groups are creating an extensive factory automation (FA) system by setting up subcommittees for laser, CAD/CAM, and LAN. Initially, the project target was development of Flexible Manufacturing System [FMS] element technology.

In 1987, another national project by MITI, the "Advanced Material-Processing and Machining System", was begun. This project will end in 1993. It centers around R&D on an excimer laser and an ion beam for the technological development of processes. These include the superprecision and super-high-quality surface treatment processes needed in such advanced-technology fields (on which Japan's future depends) as energy, precision machinery, electronics, aeronautical navigation, and aerospace. The budget figures are ¥1.10 billion (£4.23 million or $6.88 million) for 1987, ¥1.69 billion(£6.46 million or $10.49 million) for 1988, and ¥2.36 billion (£9.09 million or $14.77 million) for 1989.

The aims of this project are :

(a) to establish a research base, including a large-scale R&D facility

(b) to improve R&D in basic and advanced fields, and

(c) to strengthen international research cooperation.

In fiscal 1988, MITI added a business category to the New Energy Development Organization, and named the restructured body the New Energy and Industrial Technology Development Organization. MITI also created a research-based establishment comprising the Underground Weightless Environment Experiment Center, Research Center for the Industrial Utilization of Marine Organisms, and the Ion Engineering Center. In fiscal 1989, the Laser Application Engineering Center based on the laser group of the Industrial Research Institute (Director: T. Fujioka) joined them with a three-year ¥3 billion(£11.54 million or $18.75 million) appropriation.

One of the laser-related technologies is uranium enrichment. In January 1987, the Uranium Enrichment by Laser Technology Research Organization (UELTRO) was set up by the Electrical Enterprise Federation which consists of nine electric power companies. The new body was established to enable the private sector to accelerate laser enrichment technology development that had been achieved by the Science and Technology Agency. MITI supported the effort by appropriating ¥4.4 billion (£16.92 million or $27.50 million) for fiscal 1987. In addition, the Laser Technology Research Institute (set up by three electric power companies and six electronics firms in June 1987) is doing basic research on a new laser unit required by UELTRO. The Institute is also engaged in research activities ranging from basic laser technology to applications like precision machining and instrumentation.

In Japan, therefore, the private sector, public research agencies, and universities have all set up projects or organizations for R&D activities and are sharing of the results.

1.2 Subsidies

Major appropriations by government include the "Scientific and Technological Promotion Adjustment Working Funds" and the "Special Research Funds."

First, the former funds were established in 1981 to strengthen cooperation among private bodies universities, and governmental agencies. The funds were appropriated for research activities considered important by the government's Science and Technology Conference Policy Committee. They are used by several research institutes at MITI's Science and Technology Agency, universities, and private sector bodies to pursue "research on high-output lasers, variable-wavelength lasers, and laser-processing technology." The research encompasses the development of these laser units and as well as the development of material synthesis and processing technology using laser-wavelength selectivity. ¥282 million (£1.08 million or $1.76 million) was appropriated in fiscal 1986.

Secondly, the "Special Research Funds", the latter funds, carried out by governmental research agencies include "research on new laser technology" by the Physical and Chemical Research Institute under the direct control of the Science and Technology Agency; and "research on chemical reaction control technology by laser" by the Chemical Technological Institute of MITI.

1.3 Joint study by universities and private enterprises

As seen from Table 1.1, Osaka University's activities are remarkable regarding the development of laser generators. This is why the Laser Society and the secretariat for a national project, the "Superadvanced Process", exists in Osaka.

Table 1.1 Joint Study between Universities and Private Enterprises

Theme	University Name	In cooperation with:
Research on the development and applications of high-power single wavelength laser.	Osaka University	Canon Sales Co. Ltd.
Research on the fabrication of highly stress-resistant, dielectric, multilayer, vapor-deposited film for high-power laser.		Canon Inc.
Research on copper-vapor laser.		NEC Corporation
Software development for nuclear fusion research. This includes optimization-designing of the Cannonball target for the glass-laser, intense-light XII model.		NEC Corporation
Research on X-ray generation by short-wavelength laser and its application.		Rigaku Electric Co. Ltd.
Research on the development and application of short-wavelength laser.		Nisshin Electric Co. Ltd.
Research on performance improvement of high-frequency, short-wavelength laser.		Mitsubishi Electric Corp.
Research on development of a single-crystal molybdenum mirror.		Nippon Mining Co. Ltd.
Research on high-power glass laser.		Hoya Corp.
Research and development of high-power excimer laser.		Shin Meiwa Industry Co. Ltd.
Research on performance improvement of high-power, high-frequency, solid-state laser.		Matsushita Electric Industrial Co. Ltd.
Research on lengthening the life expectancy of a high-frequency flash lamp.		Ushio Inc.

(Continued)

Theme	University Name	In cooperation with:
Creation of an amorphous surface by laser irradiation and its physical properties.	Nagoya University	Nogaya City
Development of a metallic-vapor laser.	Ibaraki University	Power Reactor and Nuclear Fuel Development Corporation (PNC)
Development of a semiconductor process using laser-beam chemical reaction.	Hiroshima University	Babcock Hitachi K.K.
Research on performance improvement for axial-flow laser oscillator.	Kyushu Institute of Technology	Yawata Electric Machinery Mfg.
Research on putting laser optical parts to practical use.	Takamatsu National College of Technology	Tokyo Shinku Kato
Research on laser excited emission spectro-chemical analysis of molten iron.	Okayama University	Osaka Sanso Kogyo Ltd.

2. Activities by Societies

1) The Laser Society (Dr. S. Namba, professor at Osaka University and the representative of the Laser Society, 2-1 Yamadaoka, Suita, Osaka 565, Tel.06-877-5111, Fax. 06-876-6484)

The Laser Society was founded in 1973 under the appellation of the Laser Technology Social Gathering. Its purpose was the exchange of information on laser research. The body became the Laser Society (an incorporated association) in 1978.

The Laser Society has a number of sections. It consists of the Medical Laser Subcommittee, the Laser Nuclear Fusion Subcommittee, the Laser Industrial Applications Subcommittee, and the Laser Subcommittee's X-Ray Application Technology section. Membership is about 2,000 individuals and 120 corporations, principally electrical manufacturers.

2) Laser Association (Dr. H. Kawasumi, professor of the Physical and Engineering Dept. of Chuo University and the representative of the Laser Association, 1-13-27 Kasuga, Bunkyo-ku, Tokyo 112, Tel.03-817-1737, Fax.03-817-1703)

The Laser Process Research Society was jointly founded in 1972 by laser manufacturers, import trading companies, users, and scholars to advance and popularize laser processing technology. The society's activities were extended to cover the instrumentation field and the body changed its name to the Laser Applications Technological Research Society in 1977. That name, in turn, was changed in 1982 to the Laser Association to cover laser-application technology in general.

The Laser Association holds lecture meetings, offers practical seminars, plans field trips, and issues bulletins. It provides information for the application of laser technology to production.

As of June 1, 1988, the association had a membership of 42 corporations, 135 individuals, and 42 students. Its principal constituent companies are the Nippon Steel Corporation, NKK, Kobe Steel Ltd., Kubota Ltd., Toshiba Corp., Hitachi Ltd., Mitsubishi Electric Corp., Matsushita Research Institute Tokyo Inc., NEC Corporation, Nippon Infrared Engineering Co. Ltd., Nippon Laser Co. Ltd., and Fuji Photo Film Co. Ltd.

3) Applied Physics Society (T. Sugano, professor at Tokyo University and a representative of the Applied Physics Society, Kunimatsu Building, 1-2-6 Kudankita, Chiyoda-ku, Tokyo 102, Tel.03-238-1041, Fax.03-221-6245)

The Laser Society and the Laser Association are information exchange centers for relatively practical matters, whereas the Applied Physics Society discusses the results of basic research.

The Applied Physics Society was founded in 1932 as a social body. In 1946 it was authorized as an incorporated association to study and expedite the results of research on applied physics. The Society has some 1,200 regular and 1,250 student members. Laser-related topics presented at the annual spring and autumn conventions have included quantum electronics (excimer laser, solid-state laser, dye laser, semiconductor laser basics, laser spectrum, laser control, isotope separation, and laser application), as well as optical electronics.

The number of dissertations given at the society's conventions are increasing. There were 170 dissertations in autumn 1987 and 220 in spring 1988. These are a growing number of reports on the practical implementation of laser technology. Dissertations are given mainly by the Laser Research Institute of Osaka University, the Electrotechnical Laboratory, Mitsubishi Electric Corp., and Toshiba Corp.

New trends are seen in laser oscillation applications, ranging from ultraviolet rays/extreme ultraviolet rays in a vacuum to X-ray; the free electron laser as variable-wavelength laser (research by the Electrotechnical Laboratory using accumulation rings); research by the Institute of Physical and Chemical Research (Riken) and Osaka University using high-current electron beams; and research by Nihon University using a microtron accelerator. The foregoing research endeavors are funded from the Scientific and Technological Promotion Adjustment Work Fund. Activity is also under way regarding photo-chemical reaction using short-wavelength lasers (National Chemical Laboratory for Industry, Agency of Industrial Science and Technology).

3. Governmental Scientific and Technological Policies

Japan's laser technology development process will now be described in terms of the history of technological research and governmental policies.

3.1 Large-scale research and development projects by MITI's Agency of Industrial Science and Technology

1) Flexible Manufacturing System Complex provided with Laser

In the limited-variation, mass production field represented by automobiles and home electrical appliances, an efficient, automated production system is already in place. But in small-lot production that occupies 70 percent of Japan's machine output, rationalization and labor economy are difficult to achieve, and production depends upon skilled workers. In the future, manufactures will become increasingly diversified and opportunities will expand for the utilization of new materials developed to enhance product functions.

After consideration of these factors, the Agency of Industrial Science and Technology in 1977 established the Engineering Research Association for the "Flexible Manufacturing System Complex provided with Laser." The agency also continued its research activities on development of a technology that can flexibly and quickly produce small lots of mechanical parts from metallic materials in total system. The system was successfully completed in 1985. The total expenses were ¥13.7 billion(£52.69 million or $85.63 million).

The research results were reported at ICALEO '84 (International Congress on the Application of Lasers and Electro-Optics, Nov. 12-15, 1984, Boston). The report encompassed the entire system--total system design, the high-power CO_2 laser oscillator, and processing and manufacturing technology for key optical parts.

a) R&D targets (see Table 3.1 and Fig. 3.1).

b) Organizations participating

Organizations related to the Agency of Industrial Science and Technology;

Mechanical Engineering Laboratory

Electrotechnical Laboratory

Government Industrial Research Institute, Kyushu

Private enterprise;

Aida Engineering Ltd.

Fujitsu Fanac Ltd.

Hitachi Seiki Co. Ltd.

Horiba Ltd.

Ishikawajima-Harima Heavy Industries Co. Ltd,

Kobe Steel Ltd.

Makino Milling Machine Co. Ltd.

Matsushita Reseach Institute Tokyo Inc.

Mitsubishi Electric Corp.

Mitsubishi Heavy Industries Ltd.

NEC Corporation

New Japan Machine Works Ltd.

Oki Electric Industry Co. Ltd.

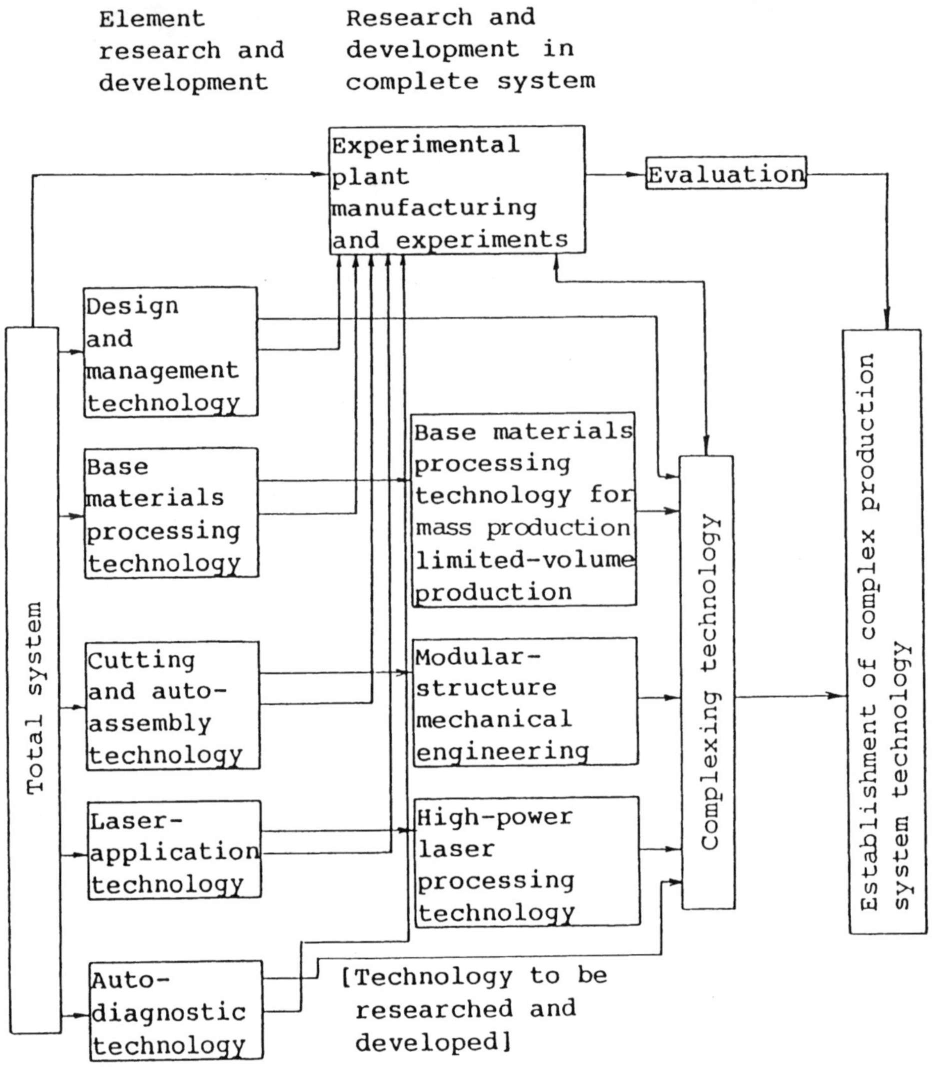

Fig. 3.1 Flow Chart for Research and Development of a Laser-based Flexible Manufacturing System

Table 3.1 Agency of Industrial Science and Technology's Targets for Research and Development in Laser-based Flexible Manufacturing Systems

Item	Target
Object to be produced	A machine constituent whose max. weight is 500kg, max. outer dimension is 1,000mm, and max. parts count is 300 (e.g., power-transfer equipment such as a gearbox, or hydraulic equipment such as a hydraulic valve).
Hours required for production	50 percent or less of the present production system.
Number of processes required for production	60 percent or less of the present production system.
High-power laser oscillator	CO_2 gas laser oscillator with an output of 20kW, an oscillating-tube life expectancy of 2,000 hours or longer, and an excited energy efficiency of 15 percent or greater.
Medium-power laser oscillator	A diluted-gas laser ascillator with an output of 200W and a solid laser oscillator with an output of 300W.

Okuma Machinery works Ltd.

Sumitomo Electric Industries Ltd.

Toshiba Corp.

Toshiba Machine Co. Ltd.

Toyoda Machine Works Ltd.

Yamazaki Iron Works Ltd.

Yasukawa Electric Mfg. Co. Ltd.

c) Outline of technological development

The laser-related technology created thus far includes development technology for a high-power CO_2 laser that performs welding and surface-hardening processing, used in conjunction with the cutting mechanism. Other welding processing and surface-hardening processing technologies have also been perfected. (These were developed by the Electrotechnical Laboratory, Mitsubishi Electric Corp., the Applied Equipment Laboratory, and Toshiba Corp.)

Research was conducted principally at the experimental plant constructed at Tsukuba, but development of the 20 kW CO_2 laser was done by the Mitsubishi Electric Corp.

An outline of the functions and layout of the Tsukuba experimental plant is provided below (see Fig. 3.2 and Fig. 3.3).

Mitsubishi Electric Corp. is evaluating low-distortion, fast welding, and surface modification processing using the 20 kW CO_2 laser. This Silent-Discharge-Assisted DC-Glow-Discharge Excitation laser (SAGE, see Fig.3.4) accomplishes gas sealing by means of a solid transmission window for laser beam discharge. (Patent unexamined 58-21891/2, 58-141887 [Mitsubishi Electric Corp., Applied Equipment Institute].)

Sumitomo Electric Industries, Horiba Ltd., and the Matsushita Research Institute Tokyo Inc. developed parts supporting the test (see Fig. 3.4).

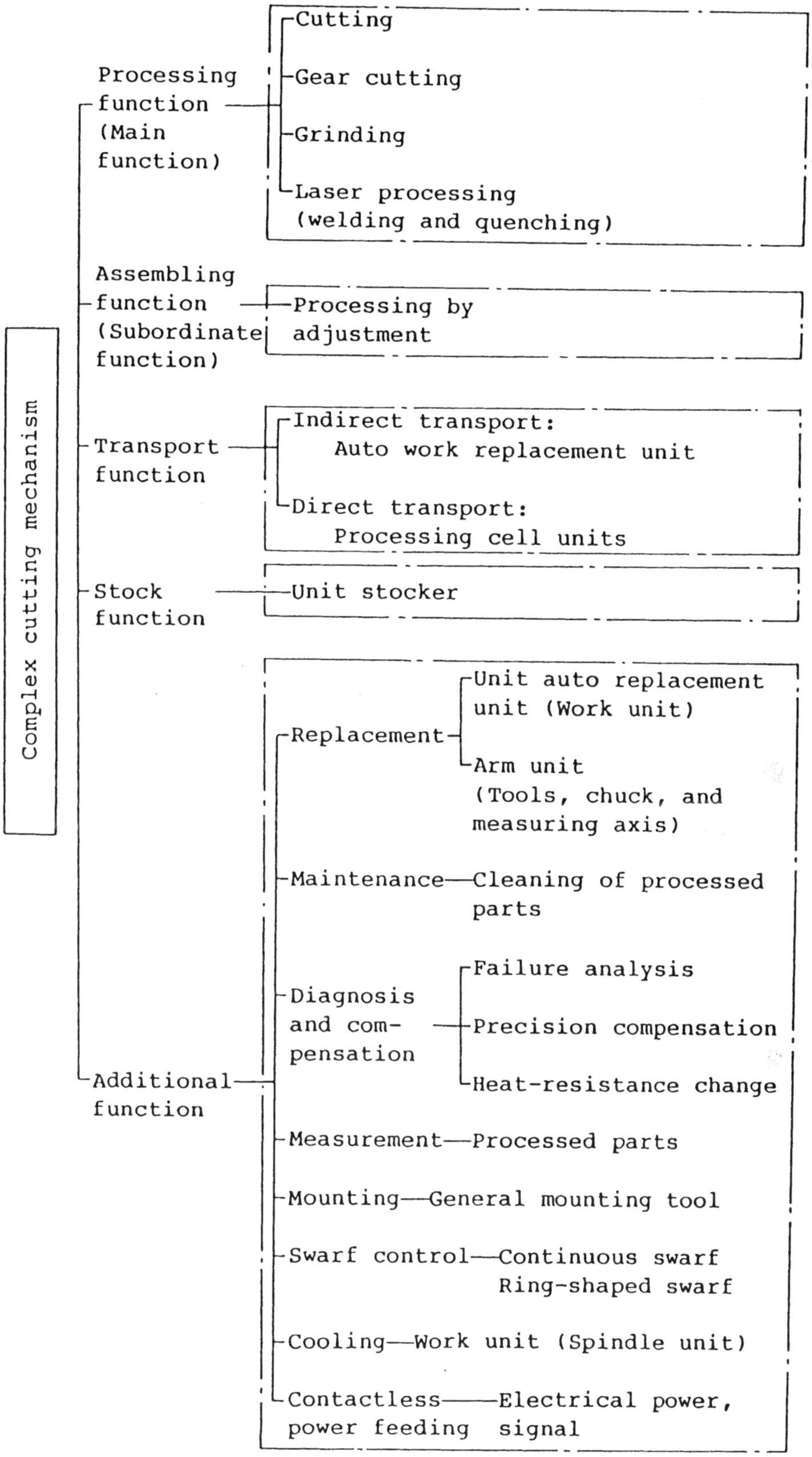

Fig. 3.2 Functions of the Complex Cutting Mechanism

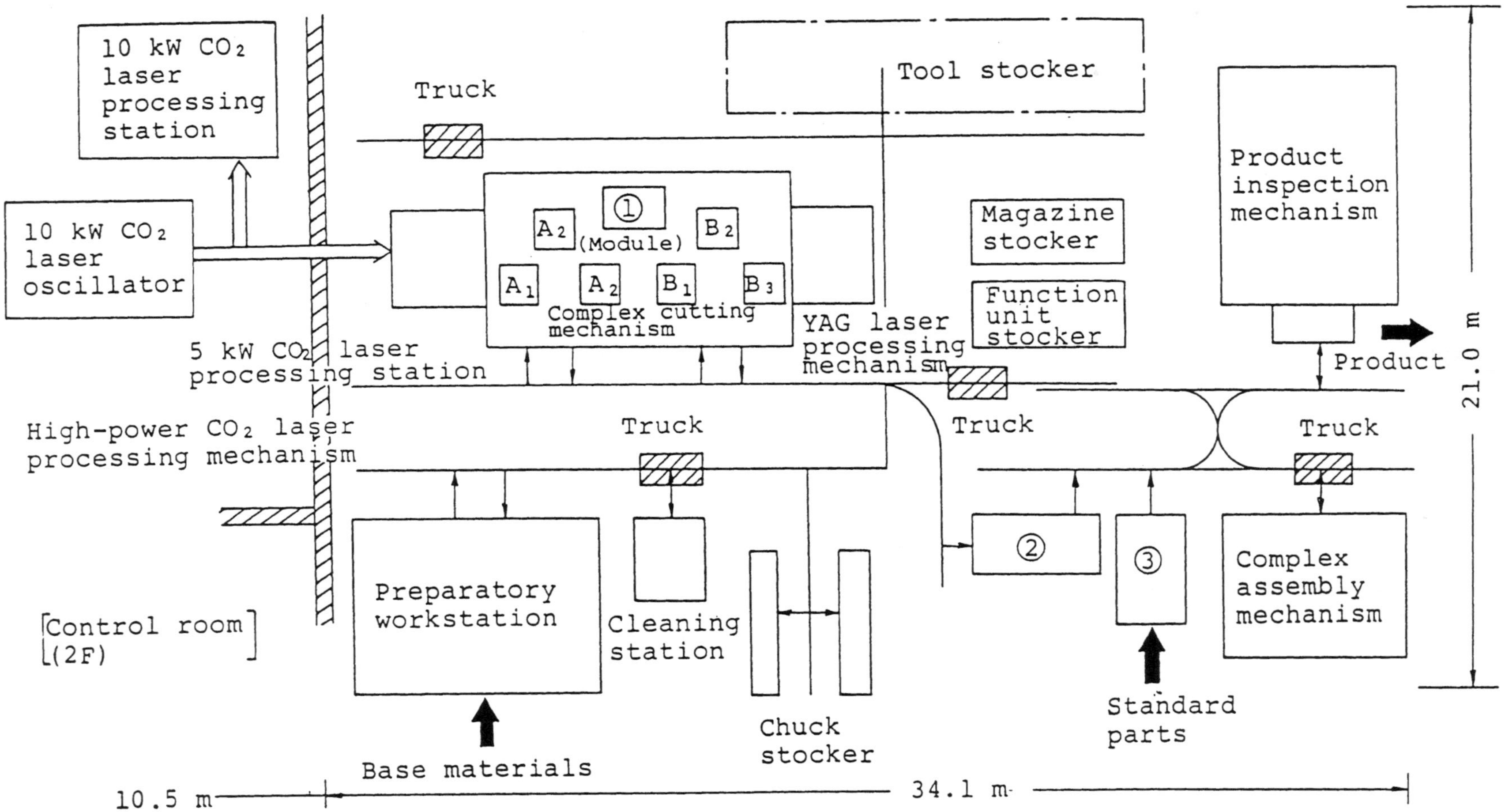

① Measuring
② Processed parts transfer unit
③ Standard parts transfer unit

Fig. 3.3 Layout of the Tsukuba Experiment Plant

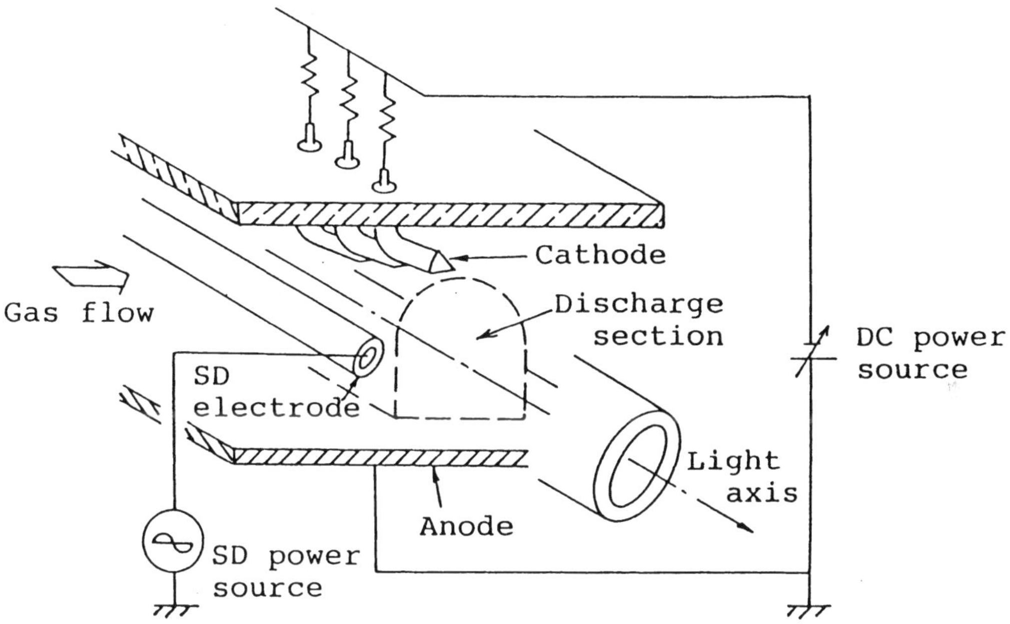

Fig. 3.4 Electrode Structure of SAGE System

In addition, NEC Corp. and others were working on the development of a 300 W YAG laser. They also conducted research on such peripheral laser technology as the removal of burrs generated during gear-cutting and the continuous cutting of chips. (Patent unexamined 60-82283 [NEC Corp., Toshiba Corp.].)

d) Research on materials

Mitsubishi Electric Corp. studied preheating and postheating conditions on the welding process.

Toshiba Corp. studied laser surface hardening conditions, with a rastered beam and used the method of applying graphite mixed with water to help absorb the laser beam.

2) Advanced Materials Processing and Machining System

The twenty-first century will demand subnanometer processing technology base for precision machines and the electronics industries. To advance R&D systematically and efficiently in excitation technologies such as excimer lasers and ion beams (which have been developed over the last ten years in the fields of machinery, electricity, and chemicals), a national project (Advanced Materials Processing and Machining Technology Research Organization) was set up in 1986 in Osaka by the Agency of Industrial Science and Technology. This body is the first of its kind in Osaka.

Since the project is ongoing, and since it is a typical national project, we will describe it in some detail.

The organization is described in Fig. 3.5

The Advanced Materials Processing and Machining Technology Research Organization was established to conduct research and development on a total system to perform superprecise, supermicro processing and superhigh-quality surface modification processing--feats that have never been achieved by existing processing technology. R&D will be conducted using high-density,

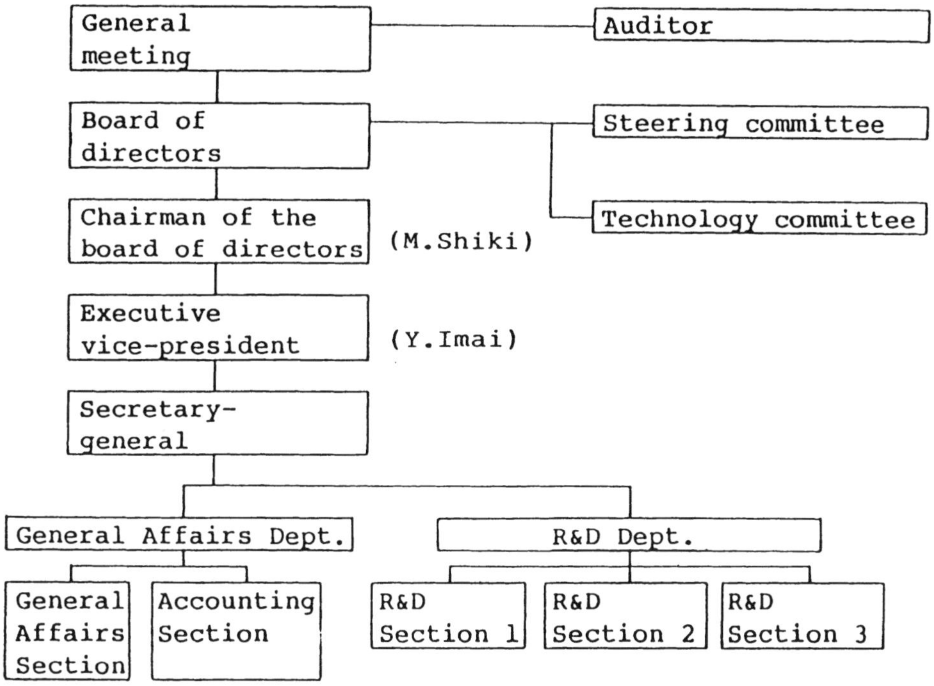

Address of Secretariat: Osaka Castle Hotel, 4F,
2-35-7 Kyobashi, Higashi-ku, Osaka

Fig.3.5 Advanced Materials Processing and Mashining Technology Research Organization high-energy-beam technology such as excimer lasers, ion beams, and superprecise machine processing units. The budget is approximately ¥15 billion(£57.69 million or $93.75 million).

Research and development items, description, and schedule are gigen in Table 3.2

i) Processing technology using excimer or ion beams

< High power excimer laser technology >

Because the excimer laser can emit ultraviolet rays, its photon energy is so high that it decomposes not only molecules, but also reactive gas on the surface of an object. Furthermore, this laser controls chemical reaction at the quantum level, thus performing supermicro processing, laminating different materials, and surface treatment for objects previously impervious to such treatment. Fig. 3.6 shows high power excimer laser of short wavelength and high frequency.

Diverse light-excitation processes can be performed by the short wavelength, high frequency, high power excimer laser. This laser has been realized by the development of key technology as below:

< Key technology >

* Rapid-charging technology
* High-efficiency discharge excitation technology
* High-speed circulation and cooling technology for corrosive laser gas
* Spare ionization source technology
* Long-life discharge structure technology
* laser beam restoration technology

These achievements have made possible that the creation of synthesis-resistant materials, the formation of high quality thin films, and supermicro processing.

Table 3.2 Research and Development Items and Schedule for the Advanced Materials Processing and Machining Technology Research Organization

	Research & development item		Description	1986	1987	1988	1989	1990	1991	1992	1993	
Entrust-ed test and research	Research and development of super-advanced processing unit technology	①Excimer laser technology	Development of short wavelength excimer laser technology and high output high frequency excimer laser technology for upgrading materials' surface quality.	Investigation and planning	Development of element technology		Manufacturing prototypes and conducting experiments	Interim evaluation	Manufacturing	Improvement	Test	Overall evaluation
		②Ion beam technology	Development of high-forcusing, wide-energy ion beam technology and high current high energy ion beam technology for fast modification of material surface layer and upgrading of the surface function.	Investigation and planning	Development of element technology		Manufacturing prototypes and conducting experiments		Manufacturing	Improvement	Test	
		③Super-precise machining technology	Development of superprecision machine processing unit technology that can process components of three-dimensional, complex shape in a highly efficient and precise manner.	Investigation and planning	Development of element technology		Manufacturing prototypes and conducting experiments		Manufacturing	Improvement	Test	
	Research and development on superadvanced processing technology		Development of superadvanced processing technology using excimer lasers, ion beams, and superprecise machine processing units.	Investigation and planning	Development of element technology		Optimization test		Manufacturing prototypes and conducting experiments	Improvement	Test	
	Research and development on supporting technology		Research and development of superprecise measuring evaluation technology supports superadvanced processing.	Investigation and planning	Basic design				Detail design	Manufacturing	Test	
	Research and development on total sytem technology		Conceptual design for construction of the superadvanced processing system and comprehensive test for the superadvanced processing system.	Investigation and planning	Conceptual design				Comprehensive test			
General test and research	Investigation into technology trends		Investigation into domestic and foreign technology trends for the superadvanced processing system, plus the technology forecast.	Investigation and planning	Investigation into technology trends							
	Investigation into the current conditions of domestic and foreign agencies		Investigation into the actual superadvanced processing system conditions at the national and public research agencies and private enterprises in Japnan and abroad.	Investigation and planning	Investigation into actual conditions of domestic and foreign agencies							

Total amount of budget : About ¥15,000 million (£57.69 million, $93.75 million)

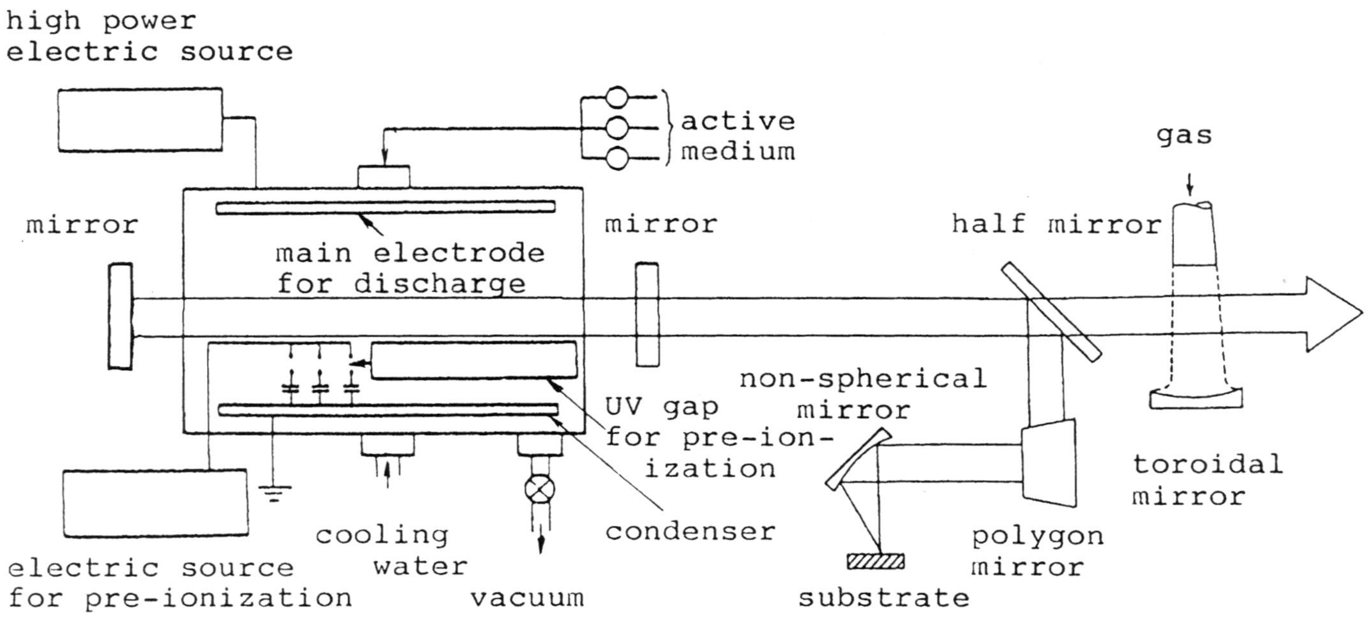

Fig. 3.6 Concept of high power excimer laser

The Members of the Advanced Materials Processing and Machining Technology Organization are as follows:

Canon Inc.

Hitachi Ltd.

Industrial Research Institute

Japan Aviation Electronics Industry Ltd.

Japan Fine Ceramics Center (JFCC)

JEOL Ltd.

Kobe Steel Ltd.

Komatsu Ltd.

Materials Process Technology Center

Matsushita Research Institute Tokyo Inc.

Mitsubishi Electric Corporation

Nachi-Fujikoshi Corporation

NEC Corporation

Nippon Kogaku K.K.

Nippon Sheet Glass Co. Ltd.

Nisshin Electric Co. Ltd.

Sumitomo Electric Industries Ltd.

Toshiba Corporation

Toyoda Machine Works Ltd.

ULVAC Corporation

Yokogawa Electric Corporation

Regarding excimer laser oscillators, the NEC Corporation, which has obtained good results in the YAG laser field, achieved an oscillation efficiency of 2.3 percent and 1,400,000 continuous oscillations. This was done by suppressing the generation of impure gas caused by reduction of the discharge time (30 nanoseconds). The suppression was achieved by using a dual-structure tube. This prevents the degradation of the circuit parts by separating the circuit from the Ar and F gas mixture.

Cannon Inc. is playing an important part in micro process research. Its laboratories have produced good results in the fields of super-high resolution, spherical lens shaping, the formation of accumulated functional films, and especially in semiconductor manufacturing units employing these technologies. Although patents from the project are unclear, 330 patents from Canon Inc. were laid open to public inspection in 1987 alone, classified as IPC-H01L21 relevant to semiconductor manufacturing units.

R and D of thin film formation and surface modification technologies are being executed by the National Chemical Laboratory for Industry in the Agency of Industrial Science and Technology, jointly with nongovernmental organizations.

These bodies performed preparatory studies for research on thin film formation using the laser-CVD method. Specifically, they studied the CVD method for metal molybdenum (Mo) as a model reaction. Molybdenum carbonyl ($Mo[CO]_6$) has a vapor pressure of about 0.1 torr at room temperature. The decomposition reaction of $Mo[CO]_6$ gas using the CO_2 gas laser is examined according to the following procedure:

The inspection quantity line for $Mo[CO]_6$ is calculated by the Fourier Transformation Infrared (FTIR) method. Because $Mo[CO]_6$ is not absorbed in the oscillation region of the CO_2 gas laser, as SF_6 molecule is used as a sensitizer agent. The details of the decomposition experiment are as follows:

① Dependency of decomposed quantity of $Mo(CO)_6$ on laser fuluence(laser energy per unit area)

② Dependency of SF_6 on pressure

③ Dependency of SF_6 on beam radius

④ Effects of other coexisting gases

ii) Technology not relevant to the laser, included in the project

< High-density ion beam technology >

An ion beam has the characteristics of being both an energy source and a material source. Ion injection by energy control can create a modified layer whose plasticity is different from that of (a) bulk materials at any depth of the surface layer of the object, (b) the multilayer structure consisting of different substances, and (c) high-functional thin films. Also, ion beams resist scattering in materials. The nature of the beams enables the achievement of a high-resolution microstructure.

As elements of the technology for high-speed, large area film-forming and superprecise microprocessing, following ion beams are required.

* A cluster ion beam, focusing gas ion beam, and high-frequency quadra-pole acceleration high energy ion beam that enables deep-layer ion injection.
* Metal ion beam
* Complex ion beam
* Superprecise processing

< Super precision machining >

Three-dimensional, freely curved surfaces at the molecular and atomic level are created with a contactless method that polishes using elastic destruction and mechanochemical reaction of the materials to be processed by the impact of abrasive grains in the processing liquid.

* Development of a structure having high rigidity and low heat deformation by employing new materials
* Development of high-rigidity cutting tools
* High-precision positioning control technology
* Contactless measuring of objects to be cut
* Processing technology using high-speed feedback control

3) A conception of Laser Application Engineering Center

Innovative laser application technology requires a large amount of trial and error to determine adequate laser wavelengths and characteristics. There are many types of laser units and most of them are extremely expensive.

Consequently. it was difficult for a single research agency to acquire all the desired lasers.

However, a plan is now under way with half funds financed by New Energy Development Organization (NEDO), that different type of the most advanced lasers will be situated at a single location and shared by a large number of researchers from the private sector.

Outline of the plan is as follows:

Initial investment (fiscal 1989-1991)

¥3,000 million (£11.54 million or $18.75 million)

Building and supplementary arrangement

¥1.2 billion(£4.16 million or $7.5 million)

Laser generator

¥1.8 billion(£6.92 million or $11.25 million)

The center prepares laser units of various wavelengths and power, including the most advanced 5 kW-grade CO laser, and CO_2, I, excimer, YAG and dye lasers. In this way the center will be ready at any time to generate whatever laser beams are required. The center is to operate in the end of 1991.

The first and second floors are occupied by laser generating rooms. The third and fourth floors are devoted to research rooms. Laser beams are carried to each research room by optical fibers or mirrors via the light transfer path situated at the center of the building.

The center handles facility maintenance, management, and research room rentals. Operational expenses are defrayed by the rental income. When a requisition is received to use a laser not yet developed, a request is made for technical guidance or for entrusted research. The center then takes appropriate action in close cooperation with the Institute of Research and Innovation.

The users include companies, universities, and national and public laboratories involved with machinery, metals, biology, and chemicals.

A report on an ideal way of R&D enforcement was published in June 1989 through MITI by the working group of the center (chairman : Dr. Ikuzo Tanaka, an ex-president of Tokyo Institute of Technology) to promote and materialize the conception of the center.

4) Uranium Enrichment Technology Employing Lasers

On January 30, 1987, nine electric companies established the Laser Enrichment Technology Research Organization. The participants are the Tokyo Electric Power Co. Inc., Japan Atomic Energy Generation Laboratory, Nihon Atomic Energy Industry Laboratory, and Power Central Laboratory.

About ¥20 billion (£76.92 million or $125 million) will be spent by March 1991 to establish the basic technology for the laser-based uranium enrichment system of the twenty-first century. The chairman of the board of directors is M. Toyoda, vice president of the Tokyo Electric Power Co. Research on the laser system, which replaces the existing gas diffusion system or centrifugal separation system, has started in the United States and France. Japan will develop the laser system independently because of the difficulty of its technological introduction.

The Research Organization produced a prototype laser enrichment unit using the presently available technology and collect the necessary basic data during fiscal 1987. In addition, the organization makes a system that can produce from 1 to 5 tons of separation work units (SWU) per year and makes a comprehensive evaluation by fiscal 1990. After fiscal 1991, demonstrative research will be carried out, including pilot tests.

To support the Research Organization, the Tokyo Electric Power Co., Kansai Electric Power Co., Chubu Electric Power Co., and six electrical manufacturers including Hitachi Ltd. and Toshiba Corp., set up the Laser Technology Laboratory at the end of June 1987. The laboratory will spend about ¥200 million(£0.77 million or $1.25 million) per year on researching the basics of laser technology, precision processing, measurement, and applications for uranium enrichment.

The laboratory has its head office in Osaka. The Chairman of the board of directors is K. Iida, vice president of the Kansai Electric Power Co. Inc. The head of the Laboratory is T. Yamanaka, professor emeritus of Osaka University. The Laboratory began its research activities in the Technology Laboratory of the Kansai Electric Power Co., in Amagasaki City, Hyogo Prefecture, with three research sections and five full-time researchers.

The first theme was the development of uranium enrichment technology employing lasers. The group was engaged in the basic research for a new type laser unit entrusted by the Laser Enrichment Research Organization.

The uranium enrichment technology employing an atomic gas laser system is being developed to separate natural uranium into U235 and U238. Power generation requires a U235 content exceeding 3 or 4 percent, but natural uranium contains only 0.7% of U235. Conventional enrichment technology uses a centrifugal separation system, but the laser system enables more efficient and economical enrichment.

The laboratory began construction of the prototype uranium enrichment laser system in the spring of 1988, and conducted tests in the summer. The tests showed that the laser output target value was achieved, thus fueling expectations for the development of a new technology.

Two types of laser are used; copper-vapor laser and dye laser. For a single laser, past experiments with copper-vapor laser showed an output of 125 W, which surpassed the target output of 100 W. For four lasers in series, past experiments showed 259 W which again exceeded the targeted output of 200 W. For dye laser output, past experiments with five lasers in series showed an output of 42 W, well beyond the target of 40 W. The future goal is to prolong life expectancy for a fixed output and improve efficiency. The life expectancy goal is 200 hours.

Toshiba Corp. engaged in the development of a high-output copper-vapor laser. This is a relevant technology in the continued development of 20 W-and

50 W-single lasers. Toshiba already achieved 81 W continuous oscillation in December 1987 by using highly-refined, pure, precision alumina tubes.

The newly developed copper-vapor laser used high voltage pulse-switching circuits and a new alumina insulator for stabilized oscillation of the laser. These are in addition to the aforementioned alumina tube of minimal impurity, which produces a structure superior in heat insulation and electrical characteristics.

Mitsubishi Electric Corporation's Application Equipment Laboratory is located in Amagasaki City, Hyogo Prefecture, and is headed by Y. Onishi. The Laboratory independently developed a high power copper-vapor laser generating system which is a core technology for uranium enrichment. This has been receiving attentions as the next generation uranium enrichment system, achieving a laser output of 102 W.

This output was achieved by improving the heatproof structure of the laser tube. A 2 MW pulse output is scheduled to be achieved in two years, thus establishing this technology for industrial use.

Regarding laser-based uranium enrichment, the Institute of Physical and Chemical Research continued research using a molecular laser enrichment system in parallel with a national project. The institute produced 3 percent enriched uranium required for atomic power generation.

The molecular laser system irradiates UF_6 (uranium hexafluoride) with infrared, which causes chemical changes only in molecules containing U235. So far, optimum laser irradiation conditions have not been obtained, and enrichment levels remain low.

Successful development of a multiwavelength laser with a wide-wavelength range (7-46 μm) by means of the induced Raman rotational scattering of CO_2 laser beams by means of para-hydrogen was a key to the 3 percent enrichment achievement. Numerical analysis using simulation models also helped in this achievement. U235 content after laser irradiation was 3.3 percent, or 4.7 times the initial concentration of 0.7 percent. This exceeds the targeted value of 3 percent.

3.2 Research and development using the science and technology promotion adjustment working fund

The science and technology promotion adjustment working fund was established in 1981 to strengthen the links between industries, universities, and governmental organizations. It is used for important research activities designated by the Science and Technology Congress Policy Committee.

Research and Development was approved for high-power, variable wavelength lasers and laser processing technology.

3.2.1 Research goal

For the high power, variable wavelength lasers that constitute the base for laser processing technology, new solid state lasers are being developed along with laser processing technology employing selected laser wavelengths (see Tables 3.3 and Table 3.4).

In the first stage of the pursuit of a new solid state laser, the formative technology for a chrysoberyl-type crystal with titanium, chromium and other materials added was established. Also studied was the possibility of the crystal being used in a new variable wavelength, solid laser. Another achievement was the slab-type glass laser, the technology for obtaining high-power base waves for variable wavelength lasers.

Theoretical research was done on oscillation mechanisms and the basic oscillation technology was developed for free electron lasers.

In laser processing technology, film formation technology for semiconductors and metal was developed along with the technology for synthesis. The materials processing technology was also developed, including the addition of impurities.

Table 3.3 Annual Planning under the Science and Technology Promotion Adjustment Working Fund

Research Item	FY 1984	FY 1985	FY 1986	FY 1987	FY 1988
(1) Development of high-output, variable-wavelength lasers					
① Development of variable-wavelength, new solid-laser oscillation technology					
i) Three-component oxide (chrysoberyl-type) laser	Establishment of active-iron-added chrysoberyl-type crystal-growth technology and laser oscillation element technology. ←→			Advancement of titanium-added chrysoberyl-laser oscillation technology. ←→	
Garnet-type laser	Establishment of active-ion-added garnet-type crystal-growth technoloty. ←→			Completion	
ii) Glass laser	Manufacturing of active-ion-added laser glass and establishment of a high-output, variable-wavelength oscillation technology using slab-type laser glass. ←→			•Advancement of high-output, slab-type, variable-wavelength laser-development technology. ←→	
② Development of the free-electron laser oscillation technology					
i) Free-electron laser using storage rings	Establishment of basic technology for visible-range free-electron laser oscillation ←→			Establishment of visible range free-electron, laser-oscillation technology ←→	
ii) Free-electron laser using high-current electron beams	Improvement of high-current electron-beam source, clarifying the interaction between electron beam and wiggler, and establishment of the element technology.			Establishment of wavelength transformation technology for the free-electron laser using an induction accelerator and improvement of oscillation technology	

(Continued)

Research Item	FY 1984	FY 1985	FY 1986	FY 1987	FY 1988
iii) Development of a small accelerator for free-electron lasers	Manufacturing of injectors, accelerating tubes, electoro-magnets, small accelerator, and other items.			Establishment of small accelerator technology for free-electron lasers.	
(2) Development of materials-synthesis technology using high-power, variable-wavelength lasers Development of film-forming technology		Selection of reactive gas and optimization of the reaction conditions.			
① Development of semiconductor film structural-control technology				Development of semiconductor surface film structural-control technology.	
② Development of metal film-formation technology				Development of a selective growth system for metal films.	
Development of an impurity addition technology		Selection of reactive gas and optimization of the reaction conditions.		(Completion)	
Development of reaction-process measuring and evaluation technology		Establishment of detection technology for the density of unstable chemical species, internal energy, etc.		(Completion)	
(3) Research promotion					
Total amount (million)	¥343 (£1.32 or $2.14)	¥415 (£1.62 or $2.59)	¥282 (£1.08 or $1.76)	¥309 (£1.18 or $1.93)	

Table 3.4 Research Items and Expenses under the Science and Technology Promotion Adjustment Working Fund in Fiscal 1987

Research Item	Organization	Expenses (million yen)
(1) Development of the high-power, variable-wavelength laser		267.4 (£1.3 million) ($1.67million)
① Development of variable-wavelength, new solid-laser oscillation technology		
i) Three-dimensional oxide (chrysoberyl-type) laser		
a) Crystal-growth technology	Mitsui Mining and Smelting Co. Ltd.	
b) Laser-oscillation technology	Institute of Physical and Chemical Research (Riken)	
ii) Glass laser		
a) Oscillation technology	The University of Tokyo	
b) Non-linear optical-crystal-growth technology	Asukaru	
② Development of free-electron laser oscillation technology		
i) Free-electron laser using accumulator rings	Electrotechnical Laboratory of MITI's Agency of Industrial Science and Technology	
ii) Free-electron laser using high-current electron beams		
a) Development of wavelength-change technology	Institute of Physical and Chemical Research (Riken)	
b) Development of oscillation technology	Osaka University	
iii) Development of a small accelerator for free-electron laser	Nihon University	
(2) Development of materials-synthesis processing technology using high-output, variable-wavelength lasers		41.5 (£0.16million) ($0.26million)
① Development of semiconductor film-structural-control technology	Institute of Physical and Chemical Research (Riken)	
② Development of metal-film formation technology	Central Laboratory of Hitachi Ltd.	
(3) Research promotion	Research and Development Bureau of the Science and Technology Agency	0.25 (£0.96thousand) ($1.56thousand)
Total		309.15

The second stage is based on the research results of the first stage and covers the following:

Regarding chrysoberyl-type crystal, research and development will mainly be done on titanium-added crystal. For the slab-type glass laser, advancement of high power, variable wavelength technology will be carried out. Establishment of oscillation technology is planned for free-electron lasers.

In laser processing technology, research is to be done on wavelength dependence of growth, mainly of semiconductor and metal films, as well as on the selective film-growth process.

3.2.2 Outline of research

(1) Development of the high power, variable wavelength laser

a) Development of a new variable wavelength, solid state laser

In three-component oxide lasers, the second stage will conduct research on high quality, large diameter, crystal growth technology and the variable wavelength oscillation technology for titanium-added chrysoberyl that achieved fixed-wavelength laser oscillation in the first stage. The goal is a high power, variable wavelength laser in the range of 700 to 1000 nm.

For glass lasers, the first stage developed the slab-type laser which promises a high power system both theoretically and experimentally. The first stage also saw the establishment of the technology to produce a variable wavelength, high power basic wave laser. An output of 200 kW was achieved. The second stage will advance the high power technology, to obtain a high kW-grade output of a fundamental wave, and develop a high power, variable wavelength laser in the infrared to ultraviolet range. Higher harmonics will be generated and parametric oscillation will be employed, using large diameter nonlinear crystal optics.

b) Development of free-electron laser oscillation technology

For the free-electron laser using storage rings, the first stage produced interference-radiant rays in the range of green to red. This was accomplished by introducing electron beams from storage rings. The second stage will establish visible range, free-electron laser oscillation technology, based on the results of the first stage.

For the free-electron laser using high current electron beams, the first stage produced good high current, relativistic electron beams and a multistage induction accelerator for the acceleration of high current electron beams. The second stage will conduct research on advancement of the free-electron laser by the use of free-electron laser wavelength conversion technology with better electron beam sources, to develop the basic technology required for future development.

In the development of a small-scale accelerator for the free-electron laser, the second stage, based on the major elements completed in the first stage, will establish technology that can generate good electron beams with narrow energy expansion, and homogeneous direction. The target is the development and evaluation of an accelerator with an acceleration energy of 35 MeV.

Other R&D on free-electron laser is being conducted by the Institute of Laser Engineering in Osaka University, although it gets behind in the R&D of other nations.

(2) Development of the materials synthesis processing technology using the high-power, variable wavelength laser

The first stage (a) developed the formation technology of silicon and tungsten thin films using the excimer laser, (b) clarified the relationships between the film formation speed and experimental conditions, and (c) studied the film formation mechanism. The second stage, based on the results obtained in the first stage, will conduct research on peculiarities of the process at different wavelengths. Studies will also be carried out on the use of the high-power, variable wavelength laser to be developed in this research.

To facilitate these studies, research will be conducted on the direct formation of semiconductor thin films by pattern projection. Multilayer film formation research will follow on the surface-structure control system. For metal thin films, research will be done on selective growth by partial modification of the substrate surface state.

Free-electron laser was adopted as one of joint R&D subjects in Kinki District, and its R&D scheme is being carried forward by Dr. Chiyoe Yamanaka, an emeritus professor of Osaka University, as the central figure. A survey was published by Kansai Kasseika Center in March 1989.

3.3 Special research funding

The status of research financed by the special funds for research agencies under the direct control of the government, one of the pillars of the government's research promotional activities, is described below.

3.3.1 Research on control technology of the laser chemical reaction by the Chemical and Technology Laboratory

(1) Goal

This research centers around a search for a highly selective laser reaction of high yield that cannot be achieved by conventional chemical reaction methods. This search is based on the high intensity and monochromaticity that characterize laser beams. The project will attempt (a) to clarify the basic laser chemical reaction process, (b) ensure that laser technology will provide acceptable reaction control, and (c) contribute to the establishment of the bases of laser technology development in the chemical industry.

(2) Contents of the research

a) Effects of the container wall's surface (on which disparities were assumed between reaction driven by laser and by heat) were studied by taking aliphatic amine and alcohol as examples. The examination proved that highly selective laser beam reaction is obtained in a path with a very low activation energy. Also examined were the creation of an active intermediate product, such as carbene radical, and the creation process, using the time-decomposition spectrum technique.

Regarding the laser-induced chain reaction, the allyl chloride laser process involving the dehydrochlorination of 1, 2-dichloro propane was measured against temperature, laser-beam flux, and the nature of the reaction container. The target allyl chloride was selectively produced with 50 percent to 60 percent purity, and expectations were heightened for a low-temperature process.

Also synthesized was a platinum ligand complex compound, $PtCl (SnCl_3) (NH_2 C_6H_4Cl)(PEt_3)$. With using it as catalyst, light activation was attempted for a hydrogenation reaction to an olefin, and a long wavelength of 300 nm or above achieved hydrogenation of the olefin.

b) Experiments will be conducted on the laser-induced reaction of mixed reactants, and characteristics of the reaction process of more than two kinds of molecule will be examined. Reactions between compounds having a great absorption strength will also be examined. These include SF_6, BCl_3, C_6, F_6, and elemental molecules such as H_2 CH_4, N_2, adn O_2. Studies will also be made on the sensitizing role of compounds having great absorption strength.

Another task will be an examination of reaction control achieved by the effects of admixture with inert gases. This is a reaction-control technique for the excitation-activation species generated by laser beams.

(c) Singular laser reaction

A search will be conducted for reactions that cannot be caused by an ordinary thermal reaction. The search will employ a pulsed thermal reaction resulting from CO_2 laser irradiation. Active catalysts with unsaturated ligands were generated by excimer laser irradiation, and the conversion reaction of a hydrocarbon to a compound containing oxygen was attempted in the presence of oxygen, and a search was made for a homogeneous catalytic reaction of high quantum yield.

It became clear to some extent that the temperature attained by gas molecules under laser beam irradiation is a function of the molecular absorption coefficient, laser power, and pressure. Observation employing a compound also revealed a tendency whereby the obtained temperature depends on total pressure. From these, the molecular heating conditions were examined using a CO_2 pulse laser.

The possibility of a laser polymerization reaction was examined by controlling the intermediate product created by excimer laser irradiation. An aromatic azide compound was used as the raw material.

The carbene (CF_2) process employing CO_2 laser irradiation was observed using the time-resolved semiconductor laser infrared spectroscopy technique. Details were obtained regarding the reaction process and the quantity reacted, and it became clear that a more active species than had been expected was produced. Also examined was the reaction of amine and alcohol to the thermal-decomposition laser reaction, along with the effects on the reaction of the container wall.

3.3.2 Research on new laser technology by the Institute of Physical and Chemical Research

Research is proceeding on the development of a coherent light source of extremely short wavelength . Other work is under way to develop a variable wavelength infrared laser. The infrared laser would facilitate the development of a laser light source exceeding the strength of existing lasers.

In the quest for a short-wavelength laser, Q-switch YAG laser beams are focused on Mg and C targets, and the stability of the created plasmas is examined. Also, a glass laser unit is constructed as a high-output laser, with the enhanced stability required for research. Achieved so far is laser energy of 10J or greater, with a pulse width of 5 ns and a stability of ±5 percent.

3.4 Laser research by the Institute of Physical and Chemical Research

To establish a foundation for Japanese laser science and technology, the laser science research group engaged mainly in studies on laser induction chemistry set up by the Institute of Physical and Chemical Research. The institute is under the direct control of the Science and Technology Agency.

3.4.1 Members of the Laser Science Research Group

Note: * Research scientist

** Visiting scientist

*** Student

**** Part-time researcher

Steering Committee:

K. Toyoda (Chairman), S. Namba, K. Tanaka, T. Kasuya, Y. Sakamoto, M. Takami, K. Takeuchi, M. Aono, K. Aoyagi, and H. Tashiro

Chief Researcher:

S. Namba

<u>Heavy Metal Isotope Separation</u>

Leaders:

M. Takami

T. Majima

Y. Matsumoto

K. Tsuchida (Hitachi Ltd.)

New Lasers

Leaders:

H. Tashiro

K. Midorikawa

Y. Kawamura	(Semiconductor Lab.)
K. Nagasaka**	(Science University of Tokyo, Faculty of Science)
M. Obara**	(Keio University, Faculty of Science and Technology)
T. Kobayashi**	(Fukui University, Faculty of Engineering Technology)
S. Sato**	(Toshiba Corp., Production and Technology Research Center, Technology Institute)
A. Suda***	(Keio University, Faculty of Science and Technology)
H. Kubomura***	(Keio University, Faculty of Science and Technology)
H. Saito***	(Science University of Tokyo, Faculty of Science)
T. Tohi***	(Science University of Tokyo, Faculty of Science)
K. Suda***	(Science University of Tokyo, Faculty of Science)
M. Uchida***	(Science University of Tokyo, Faculty of Science)
T. Tanaka***	(Keio University, Faculty of Science and Technology)
M. Ohwa***	(Keio University, Faculty of Science and Technology)

Laser Processes

Leaders:

K. Takeuchi

S. Satoda

T. Oyama

S. Kato

S. Endo (Powder Technology Lab.)

S. Arai (Catalysis Lab.)

M. Kawahashi** (Saitama University, Faculty of Engineering)

S. Matsumoto** (Saitama University, Faculty of Engineering)

T. Ishii** (Science University of Tokyo, Faculty of Science)

Y. Makide*** (The University of Tokyo, Faculty of Science)

k. Tashiro* (Power Reactor and Nuclear Fuel Development Corporation)

E. Suzuki* (Power Reactor and Nuclear Fuel Development Corporation)

J. Ichikawa*** (Science University of Tokyo, Faculty of Science)

H. Nishizawa*** (Science University of Tokyo, Faculty of Science)

K. Ishida*** (Tokyo University, Faculty of Engineering)

Y. Kobori*** (Saitama University, Faculty of Engineering)

S. Kakagawa*** (Saitama University, Faculty of Engineering)

Laser Materials Science

Leaders:

Y. Aoyagi

S. Iwai

Y. Segawa

T. Hara

T. Meguro

M. Aono (Friction and Lubrication Lab.)

M. Mihara

T. Kasuya (Microwave Physics Lab.)

K. Ando (Atomic Processors Lab.)

T. Soyama (Plasma Physics Lab.)

S. Kato (Plasma Physics Lab.)

M. Hamagaki (Plasma Physics Lab.)

A. Doi** (Int. Micro Technol. Co. Ltd.)

M. Murahara** (Tokai University, Faculty of Engineering)

J. Zeng**	(Semiconductor Lab., China)
J. Yu***	(Semiconductor Lab., China)
T. Daigo***	(Fukuoka Education University)
M. Miyoshi***	(Yamaguchi Junior College of Industry)
E. Hamamura***	(The University of Tokyo, Faculty of Engineering)
N. Nagasawa***	(The University of Tokyo, Faculty of Science)
S. Komuro***	(Tokyo University, Faculty of Engineering)
A. Moritani*	(Nippon Steel Corp.)
M. Takahashi*	(Nippon Steel Corp.)
Y. Morikawa*	(Nippon Steel Corp.)
A. Den*	(Irie Engineering Lab.)
A. Sugimoto*	(Mitsui Mining and Smelting Co. Ltd.)
M. Kagami*	(Mitsubishi Rayon Co. Ltd.)
A. Matsudo*	(Rubarian)
N. Takayama*	(Tokyo Electron Ltd.)
K. Tozawa*	(Tokyo Electron Ltd.)
T. Matsuo*	(Tokyo Electron Ltd.)
H. Nagata*	(Tokyo Electron Ltd.)
J. kusano***	(Osaka University, Faculty of Engineering Science)
M. Kanazawa***	(Tokyo University, Faculty of Engineering)
T. Takeda	(Tokyo University, Faculty of Engineering)

Since 1985, the group has contributed to the laser tehchnological foundation by its research activities in the fields of heavy metal isotope separation, new lasers, laser processes, and laser molecular processing.

3.4.2 Reports

Heavy Metal Isotope Separation

Absorbed energy and light acoustic signals in SF_6-$Fe(CO)_5$ infrared ray intensification and decomposition reaction ... T. Majima, S. Ishikawa, T. Ishii, Y. Matsushita, and M. Takami

Effects of the added gas on the infrared ray intensification and decomposition reaction of transition metal carbonyl ...T. Majima, M. Kurokawa, and T. Ishii

Measuring 5 μm transient absorption in the SF_6-$Fe(CO)_5$ infrared ray intensification reaction ... T. Majima, Y. Matsumoto, and M. Takami

Infrared polyphoton decomposition reaction of trifluoromethyl benzene derivatives ... S. Ishikawa, T. Ishii, and T. Majima

Infrared polyphoton decomposition reaction of tetra-ethyl-silicon, -germanium, and -tin ... T. Majima, K. Nagahama, and T. Ishii

Infrared polyphoton decomposition reaction of cyclopentadiethyl manganese tricarbonyl ... S. Ishikawa, T. Ishii, and T. Majima

SnH_4 ν_2/ν_4 band infrared microwave double resonance and diode laser spectroscopy ... Y. Oshima, Y. Matsumoto, M. Takami, S. Yamamoto, and K. Kutsu

Theoretical calculation of SF_6 ν_s mode vibration energy ... Y. Matsumoto, and M. Takami

Infrared microwave double resonance spectroscopy by OCS diode laser... Y. Matsumoto and M. Takami

CO_2 ν_6/ν_7 band supersonic jet infrared spectroscopy ... Y. Matsumoto, T. Majima, and M. Takami

Application of slit nozzle to supersonic jet infrared spectrum
Y. Matsumoto, M. Takami

New Lasers

Measuring of energy scattering of cold relative electron beams by backward Thomson scattering ... Y. Kawamura, M. Kawai, and K. Toyada

Research on resonance-photo-excited ultraviolet CIII ion laser by CO_2 laser-generated plasmas ... H. Kubota, S. Kawakai, K. Midorikawa, H. Tashiro, and S. Namba

Development of high-repetitive TEA CO_2 laser using a magnetic pulse compression circuit ... T. Tanaka, M. Obara, K. Midorikawa, H. Tashiro, and S. Namba

Observation of relaxation time of NH_3 rotation by short pulse CO_2 laser ...
H. Saito, H. Tashiro, K. Nagasaka, K. Toyoda, and S. Namba

Laser Processes

Uranium enrichment by molecular laser process using low-temperature cells (contrast ratio and effects of exchange reaction) ... K. Takeuchi, S. Kato, T. Oyama, S. Satooka, E. Suzuki, K. Tashiro, H. Tashiro, K. Midorikawa, and S. Namba

Effects of added scavenger gas in uranium enrichment by UF_6 molecular laser process; effects on decomposition speed and separation coefficient ... S. Kato, K. Takeuchi, T. Oyama, S. Satooka, K. Midorikawa. H. Tashiro, and S. Namba

Flow in a conical supersonic nozzle, accompanied by SF_6 condensation ... T. Shiozaki and M. Kawahashi

Numerical calculation of two-dimensional viscous compressible flow in a supersonic diverging nozzle ... E. Suzuki and K. Takeuchi

Formation of Boron particles by TEA-CO_2 laser ... T. Oyama, T. Yokota, M. Kobayashi, T. Ishii, and K. Takeuchi

Dependency of pressure on infrared polyphoton dissociation of CTF_3; experiment and simulation calculation by master equation

Laser Materials Science

Optical nature of Ti_3^+ :$BeAl_2O_4$ and laser oscillation ... A. Sugimoto, Y. Segawa, B. Kim, S. Namba, K. Yamagishi, Y. Anzai, and Y. Yamaguchi

Growth of the GaAs monatomic layer by the laser MOVPE technique ... S. Iwai, T. Doi, Y. Aoyagi, and S. Namba

Growth model for laser atomic layer epitaxy of GaAs ... T. Doi, Y. Aoyagi, and S. Namba

Electrical field effects of exciter dynamics in a multiple quantum well ... M. Miyoshi, Y. Aoyagi, Y. Segawa, S. Namba, and N. Sano

Picosecond spectrum of beams emitted from GaAs superlattice ... J. Kusano, Y. Segawa, Y. Aoyagi, and S. Namba

Small-size electron-beam excitation-ion source ... N. Takayama, T. Hara, M.

Hamagaki, G. Kawamura, Y. Aoyagi, and S. Namba

Characteristics of electron-beam excitation-ion source ... H. Hamagaki, T. Hara, Y. Aoyagi, and S, Namba

Research on recoupling laser using laser-generated plasma ... T. Hara, M. Hamagaki, K. Ando, T, Sohyama, Y. Segawa, Y. Aoyaki, and S. Namba

Deposition of crystalline silicon films by means of excimer laser beams ... N. Takahashi, M. Murahara, K. Toyada, and S. Namba

Other

Excimer laser-induced chemical vapor deposition of amorphous SiC films ... T. Taguchi, M. Morikawa, Y. Yiratsuka, and K. Toyada

Note: Copies of above-mentioned reports can be obtained from "Laser Science Progress Report of IPCR (the Institute of Physical and Chemical Research), No. 9, 1987.

Some of their articles are cited as follows.

Theoretical studies in support of laser-implemented uranium-enrichment research:

Numerical Calculation of Two-Dimensional Viscous Compressible Flows in a Supersonic Diverging Nozzle

Eiji Suzuki and Kazuo Takeuchi

Supersonic diverging nozzle flows were calculated by an implicit-finite difference procedure for unsteady two-dimensional flow capable of handling arbitrary geometry through the use of general coordinate transformation. The scheme is based on the Beam-Warming delta-form approximation factorization scheme, and viscous effects are incorporated with a "thin-layer" approximation of the Navier-Stokes equations. The numerical results obtained for the 2-D inviscid and viscid supersonic diverging nozzle flows have been compared with those for the 1-D isentropic flows. Comparison of the calculated static temperature distributions show no remarkable differences except for the boundary layer region in a viscous case.

The mechanism of this-film atomic layer growth:

Growth Model for Laser Atomic-Layer Epitaxy of GaAs

Atsutoshi Doi, Yoshinobu Aoyagi, and Susumu Namba

A growth model is reported for atomic-layer epitaxy (ALE) of GaAs by pulsed laser metal organic vapor phase epitaxy. The physical processes of the model are surface reactions: the adsorption, decomposition, and desorption of trimethylgallium (TMG). No absorption is assumed for TMG on itself. The decomposition and desorption of TMG are expressed in terms of reaction times. Different decomposition-time constants are assumed for TMG absorbate on Ga and As. Calculations are performed for the cases of thermal and photochemical decomposition. It is found that an enhancement of the decomposition rate of TMG adsorbate on As, compared to that on Ga caused by laser irradiation, is the main cause for the suspension of Ga deposition at 100% coverage that is an essential part of the growth mechanism for ALE.

Actual material development cases:

Deposition of Crystalline Silicon Films by Means of Excimer Laser Beams

Nozomu Takahashi, Masataka Murahara,
Koichi Toyoda, and Susumu Namba

An ArF excimer-laser beam was divided into two beams, parallel and perpendicular to a substrate surface on which a silicon film was to be deposited. A focused parallel beam induced dual-photon decomposition of SiH_4 near the surface of a substrate to produce radicals. SiH_4 and/or Si, as a vertical beam that had low energy compared with that of a parallel one, brought about excitation at the interface between SiH_4 and substrate. We investigated the effects of the surface irradiation and temperature of the substrate on the Si-film formation. We found that the refractive index of the substrate irradiated with a vertical beam of 50 mJ/cm^2 was increased from 2.80 to 3.46 at a temperature as low as 250 ℃, indicating crystalline Si deposition at low temperatures.

Excimer Laser-Induced Chemical Vapor Deposition of Amorphous SiC Film

Toshiro Taguchi, Masato Morikawa,
Yasuyuki Hiratsuka, and Koichi Toyoda

Deposition has been demonstrated for amorphous hydrogenated silicon carbide films (a-SiC:H) on a glass substrate by photodissociation of SiH_4, Si_2H_6, and C_6H_6, using a focused ArF excimer laser. The dependence of the silicon content of the deposited a-SiC:H films on the ratio of the partial pressure of Si_2H_6 to that of C_6H_6 was experimentally determined. The optical band gap of the films determined from UV-absorption measurement depended on the silicon content and increased as the silicon content decreased.

Formation of Boron Particles by a TEA-CO_2 Laser

Toshiyuki Oyama, Chiaki Yokota, Masayoshi Kobayashi,
Tadahiro Ishii, and Kazuo Takeuchi

TEA-CO_2 laser-induced dielectric gas breakdown has been studied in the reaction of a BCl_3-H_2 mixture at a variety of pressures and fluences. Besides gaseous $BHCl_2$ and HCl, brown-black powder was produced--composed of nearly spherical amorphous boron particles, monodispersed, and very fine (0.1 μm in size). The particulate material was characterized using a scanning electron microscope, ICP atomic emission spectroscopy, and electron diffraction. The formation of the boron particles is assumed to proceed via homogeneous nucleation of boron atoms in the gas phase.

In 1988, research commenced on "laser chemistry in critical conditions", general research financed by scientific research funds provided by the Mimistry of Education.

"Development of the pattern of crystal growth technique by laser MOVPE and its application to the development of a new light functional element" is the theme for the above research, and following is its content.

GaAs epitaxial crystal growth was made using a laser MOVPE under different growth conditions. Then the pattern of crystal growth produced by laser MOVPE and the growth mechanism were examined. The results showed that the Ar ion laser irradiated decomposes organic metal on the GaAs surface, but does not decompose AsH_3. It was also found that when the atomic surface of the substrate is As, the decomposition rate achived by the organic metal laser is very high within a certain temperature range. But when the atomic surface of the substrate is Ga, almost no decomposition is seen. In addition, it was found that single atomic layer epitaxy is achieved by Ar laser irradiation with the crystal growth rate dependent on the temperature, the quantity of TMG supplied, and laser strength. It was further found that the selective reaction well model explains the results of the experiment.

Other recommended lines of research are as follows:

① Research on the Laser-Induced Surface Reaction

Research on the CO_2 laser-induced reaction mechanism on the SiO_2 particle surface was conducted with CDF_3 gas present. The research results showed the coexistence of excitation of the solid itself and excitation of absorbed species molecules in the laser-induced reaction on the solid surface. The surface analysis provided individual analyses of these reaction. Changes of reaction conditions such as laser wavelength and strength indicated that surface-reaction control is possible.

② Research on the Laser-Induced Surface Reaction (Kurata Subsidy from the Promotional Organization for Domestically Developed Technologies)

A programmable temperature-raising unit and Ar ion sputtering unit for measuring the desorption spectrum of surface-adsorbed species were made and set up in a superhigh vacuum created for solid-surface research. To perform vibrational spectral diffraction of solid-surface adsorbed species, a unit for IR-RAS (infrared reflection absorption technique) was made and samples such as L-B films were employed.

Si was deposited onto a clean Pb foil from Si_2F_6 by using the above units, and the thermal reaction and CO_2 laser induction responce of the Pb surface state were compared.

3.5 Summary

As mentioned earlier, when the preceding research by private enterprises achieve substantive advancement, national projects are established in the form of research organizations to consolidate the research and expedite technological development. In general at the end of a project lasting seven to eight years, each participating enterprise manufactures its own products based on the results obtained.

Usually, prior to the establishment of a project, an investigation committee consisting of representatives of each participating company is established by an organization representing the industry. The committee studies the appropriateness of the project by conducting feasibility studies while adjusting the budget scale with the relevant governmental sections.

4. Technological Details

Of the technologies using lasers, materials processing and the development of new materials are mainly described in relation to materials.

4.1 Present status and future trends in laser processing technology

With the realization of a stable high output power laser oscillator, research and development have become active for materials processing employing its high-energy density characteristics and high-focussing nature.

Figure 4.1 shows the relationship between laser wavelength and output. Tables 4.1 and 4.2 respectively give a brief outline of fields of application and present status, as given by H. Kawazumi and R. Matsunae.

One of the salient features of lasers is the concentration of high output into a micro area as a coherent beam. This enables not only the attainment of 10^8 to 10^9 W/cm^2 power density by continuous output, but also short-term high-energy pulse output (on the order of one millisecond for CO_2, and less than one microsecond for YAG). Such high energy is mainly used as a heat source.

Also, in an application field using the rapid cooling phenomenon from a high temperature caused by laser heating, nonequilibrium condensation and cooling are expected because cooling speed becomes very high depending on conditions. This brings the possibility of formation on the material of surface layers of different characteristics compared to the matrix (surface modification) and the creation of new materials of locally different texture.

This is considered to be one of the attractive future laser application fields.

Material processing using the features of lasers applies to a variety of fields with incomplete material processing included, and can be broadly classified into the following:

○ Precise welding and fusing using deep-penetration characteristics

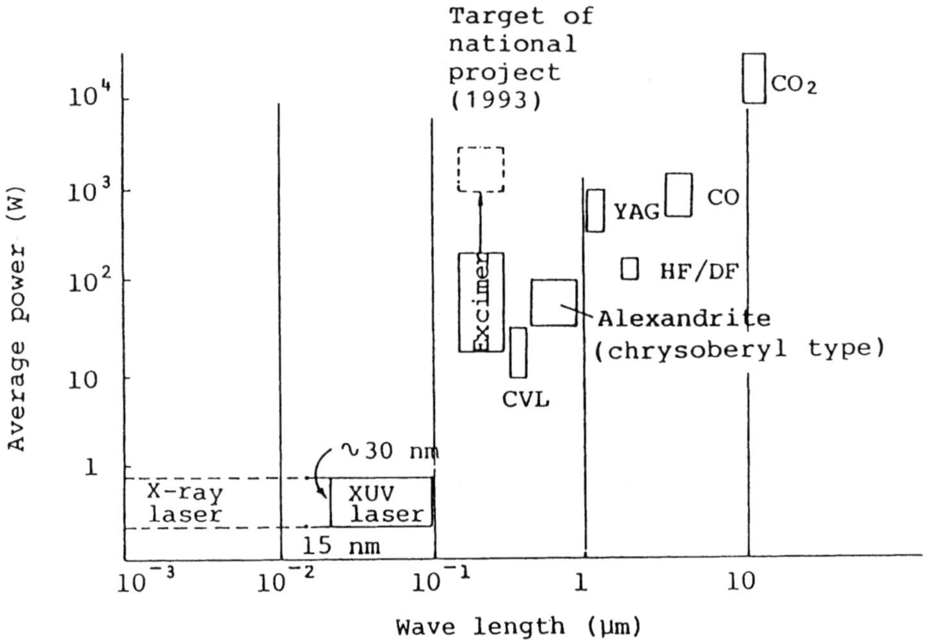

Fig. 4.1 Laser Development Status

Table 4.1 Commercial Lasers for Materials Processing

Type	Wavelength (μm)	Operational mode	Peak power (kw)	Average or CW power	Features and use
Ruby	0.6943	Normal pulse	10-100	10-100J/pulse	Not suitable for repetitive-pulse oscillation.
Glass	1.054-1.062	Normal pulse	10-100	10-100J/pulse	Perforation, spot welding.
YAG	1.064	Normal pulse	1-20	10-1000	High average power. Applicable for many uses according to operational modes.
		CW		10-1000	
		short pulse (CW-excitation Q-switch)	10-100	1-200	
Alexandrite	0.70-0.83	Normal pulse	5-50	10-100	Cu, Al drilling
CO_2	10.6, etc.	CW (closed, or low-speed gas flow)	Five times CW power at the time of pulse operation		High average power. Applicable to multipurpose processing of nonmetal and metal.
		CW (High-speed gas flow)	do.	500-20000	
		Short pulse(TEA)	5-50 MW	10-100	Marking
Excimer	0.193, 0.248 0.308 and others	Short pulse	5-30 MW	10-150	High-power ultraviolet ray source, semicounductor process
He-Ne	0.6328, and others	CW		0.5-50 mW	Measuring, optical recording
He-Cd	0.4416 0.3250	CW		1-40 mW	Short wavelength, optical recording, and exposure
Ar ion	0.5145, 0.4579, 0.35 and others	CW		0.1-5	Short wavelength, light memory recording, and exposure
Xe ion	0.5353, and others	Short pulse	0.1-1	10-100 mW	High-power, semiconductor element trimming
semi-conductor	0.77-0.88	CW		3-100 mW	Small-size, high-efficiency, optical recording

* Width of the normal pulse is 0.1-10 ms. Width of the short pulse is 1 μs or less.

Table 4.2 Type of Laser Processing and Status of Commercial Use

(1) Material removal	
a) Cutting	Practical use in the cutting of thin metal and non-metal
b) Scribing	In great demand in the semiconductor industry
c) Trimming	In great demand in the semiconductor industry
d) Marking	Rapid increase of demand
e) Drilling	Microdrilling of superhard and supersoft materials
f) Surface cleaning	Selective removal of oxidized films, and other films Technology expected to be developted in the future
g) Super-fine powder fabrication (Physical vapor deposition)	Technology expected to be developted in the future
(2) Joining (present trend shows a rapid increase in demands)	
a) Welding	Used in the mass-production of small parts
b) Brazing and Soldering	Demand is increasing in the semiconductor industry
(3) Surface modification (expansive future development is expected in this field)	
a) Surface hardening	Practical use in automobile parts, and other fields
b) Surface alloying or cladding	Technology expected to be developed in the future
c) Structure modification	Technology expected to be developed in the future
d) Laser vapor deposition	Technology expected to be developed in the future. Attractive to the semiconductor industry

○ Surface treatments employing rapid heating and rapid cooling effects--especially the latter

○ Microprocessing using microheating characteristics

i) Utilization of the deep-penetration characteristics

Relatively fundamental research has already been conducted on the precise welding and fusing technology in the fields of electron beam welding and gas welding. The research is principally concerned with problems unique to (a) the laser, (b) processed quality (defect prevention), and (c) processing software compared with other welding methods mentioned above. Among other methods, laser cutting is already in its practical use stage and replaces punch pressing in the field of sheet metal working. In these fields, the possibility is being studied for high-speed cutting of difficult-to-cut materials such as FRP. In the field of welding, research is concentrated mainly on comparison with electron beams. Specially, research is being actively conducted on the prevention of attenuation of lasers by interaction with plasmas generated from the welded part, and welding defects such as blowholes; and also on the welding of such materials as ceramics. But few of these applications are put to practical use, only high-speed welding of sheet metals is reported as a case of practical usage.

ii) Utilization of the rapid cooling effects

A great amount of research has been made, especially by American research organizations, into surface treatment using the concentrated thermal source characteristics for a variety of materials and purposes. In Japan, also, a great deal of research is under way on the same subject. Utilization of these characteristics enables a maximum cooling speed of 10^{11} K sec^{-1}, depending on the laser irradiation conditions. The following advantages are obtained.

a) Local modification of surface layers is made without causing heat-degradation of materials.

b) Heat processing is done for a precisely controlled thickness (or depth).

c) A non-thermal-equilibrium texture is obtained. That is, amorphous layers (or glassy texture), or supersaturated texture.

d) A solidified texture tends to be of a microepitaxial cell structure, and of a homogeneous texture with little component segregation.

Next, the laser irradiation process is classified into the following.

a) Processing of a solid material (hardening)

b) Processing accompanied by partial melting (annealing, glazing)

c) Introduction of additives (alloying)

But very few fields have put the characteristics to practical use. They are classified in terms of research purposes, as follows:

a) Improvement of a abrasion-resistance and erosion-resistance

b) Improvement of corrosion-resistance

c) Annealing in the semiconductor field (improving surface layer of semiconductor after ion injection, and forming fine crystal structure of semiconductor by melting with pulse laser, irradiation and consecutive rapid cooling)

d) Manufacturing materials of fine grain size

Table 4.3 shows the proportions of patents applied for in each field during the past eight years. Although it is natural that many patents commonly apply to several fields, patents are relatively few for cutting, scribing, and other fields (trimming and marking) which occupy the majority of laser application machines on the market.

This means that in these fields, lasers are relatively easy to introduce and put to practical use. On the other hand, a very large number of patents are applied for in the welding field. This means that this field attracts interest in implementing laser use, and at the same time, that a large number of

Table 4.3 Changes in Laser-Related Patent Applications by Field

% of Applications devoted to:

Year	Total No. of Applications	Welding (%)	Cutting (%)	Drilling (%)	Scribing (%)	Hardening (%)	Surface modification (%)	Others(cladding, trimming, marking) (%)	Common to processings (%)
1980	86	26	10	10	1	2	5	5	41
1981	138	38	20	7	3	3	1	4	22
1982	238	31	10	6	3	3	2	8	38
1983	290	37	12	6	2	1	4	6	31
1984	401	26	13	5	2	2	2	10	40
1985	499	23	14	5	2	2	4	11	39
1986	661	24	16	5	2	1	4	11	37
1987	675	24	14	2	1	3	5	17	34

development projects are under way. The surface modification field saw few patents despite the large number of basic research dissertations. Recent trend, however, indicates gradual increases in the fields of surface modification and cladding as shown in Figure 4.2. Most of the surface-related technologies require improvement of abrasion and erosion resistance and are given attention as future application fields.

The number of patent applications by inventive step (Figure 4.3) indicates that the number of patents applied for regarding the processing method are overwhelmingly in the majority. Note that the classification includes several patents extending over two fields, making the total number of those calculated larger than the actual total. Inventions concerning lenses and mirrors reveal a mixture of option-related (process-assisting equipment) and hardware-related patents. In this regard, option-related patents are also very numerous. As a result, the number of hardware inventions, such as laser units, is very small composed with the number of processing-software-related inventions.

Problems with individual processing technologies are taken up below, beginning with laser-cutting and the potential for laser application from a new viewpoint.

The Science and Technology Agency and Metal Material Technology Laboratory, in cooperation with the Toshiba Corp. have been experimenting and observing some items on which little data exists despite the fact that they are important issues.

4.1.1 Laser cutting

Laser cutting of a steel sheet to a thickness of several millimeters has already been put into practical use and is becoming indispensable as a tool in CAD/CAM production system. For instance, optimal cutting speeds by CO_2 laser for a variety of steel sheet thicknesses are summarized in Figure 4.4.[2] As

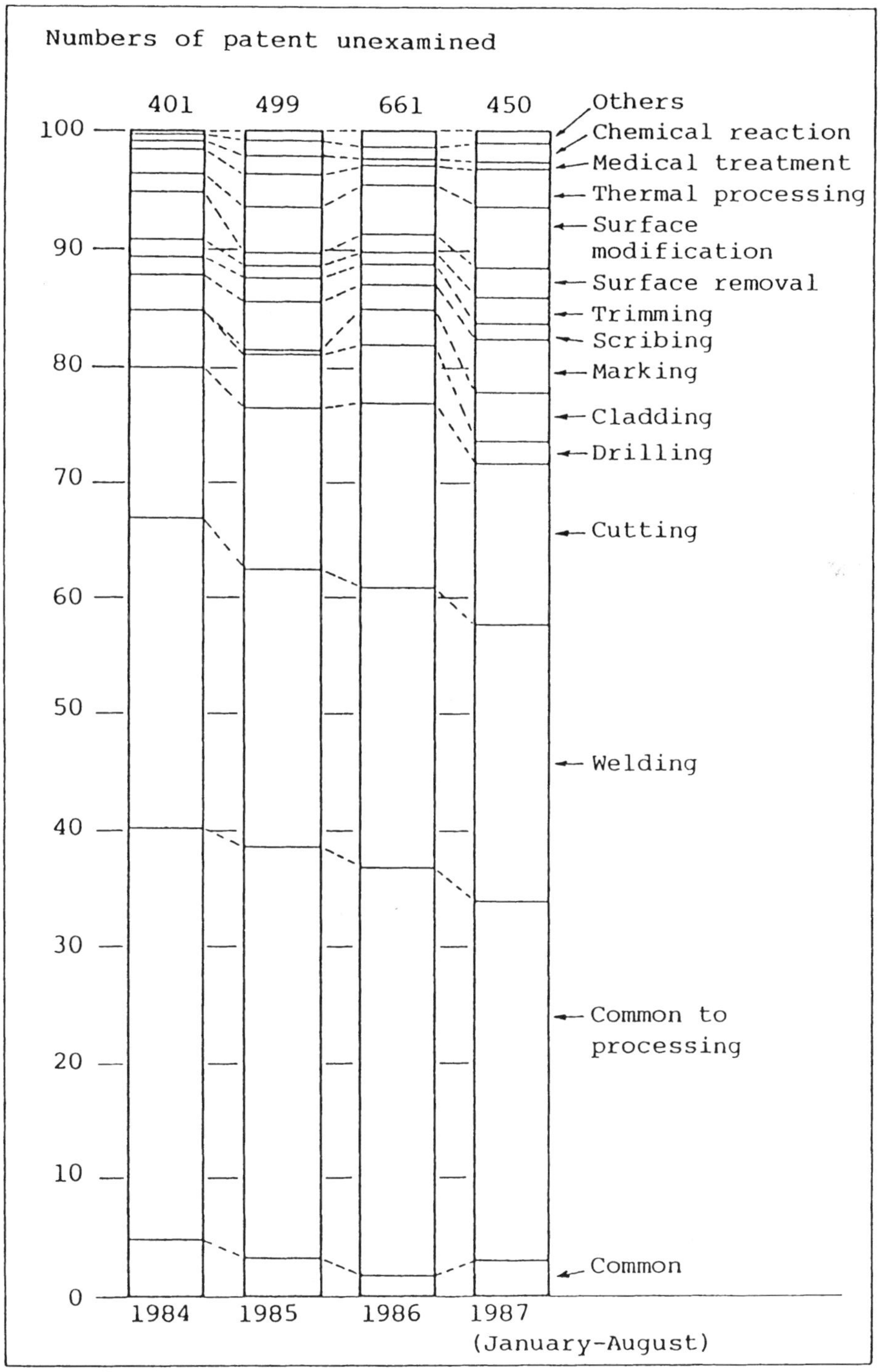

Fig. 4.2 Numbers of Laser Processing-related Patent unexamined by Inventive Fields[41)]

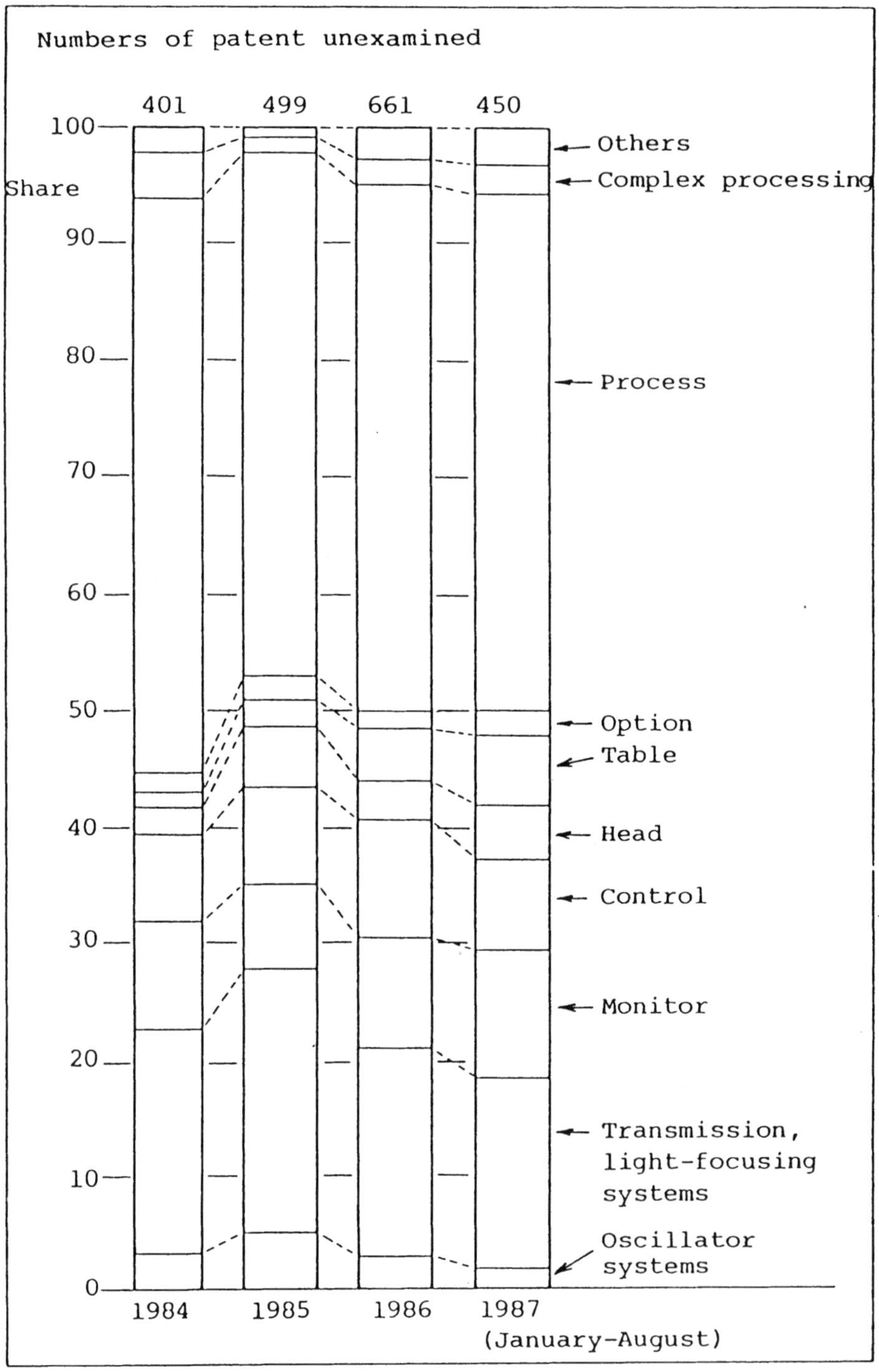

Fig. 4.3 Numbers of Laser Processing-related Patent unexamined by Inventive Step

with other fusing methods, laser cutting can be divided into laser melt cutting, laser vapor cutting, and laser oxygen cutting. Each has a different cutting quality and thermal effect.

The cutting of structual steel seems to have been put into practical use without problems up to a thickness of 10 mm or so, but when it comes to stainless steel and aluminum, the difficulty of dross-adhesion arises. The Metal Material Technology Laboratory conducted experiments on 1 mm stainless steel sheet and a colored stainless steel sheet to examine the thermal effects; and also on a vinyl-film-coated colored stainless steel sheet to examine the cutting of organic materials. A 1 kW CO_2 laser was used, with oxygen as the assistant gas. The cutting of SUS304 stainless steel sheets quickly brought about adhesion of dross, but showed particularly favorable results in comparison with other thermal cutting techniques. The same result was obtained with the colored stainless steel plate. The area in which heat-related color changes occurred in the neighborhood of the cutting part extends to about 1 mm.

A cutting quality improvement method has been proposed. With this method, dross is entirely removed by piling another material onto the back of the stainless steel sheet to be cut. Figure 4.5 shows the effects brought about when the sputtering adhesion-prevention agent whose main component is free carbon was used for this purpose.[3)]

Extreme difficulty was experienced in cutting a vinyl-film-coated stainless steel sheet. This seems to be a result of the scattering of laser beams by gas generated by film combustion.

Low-temperature experiments showed adhesion of resin residues on the surface and adhesion of dross on the back, which cannot be termed a favorable cutting process.

A laser beam was used in cutting an aluminum alloy coated with plastic sheet used in aircraft construction. Because the beam reflects off the surface

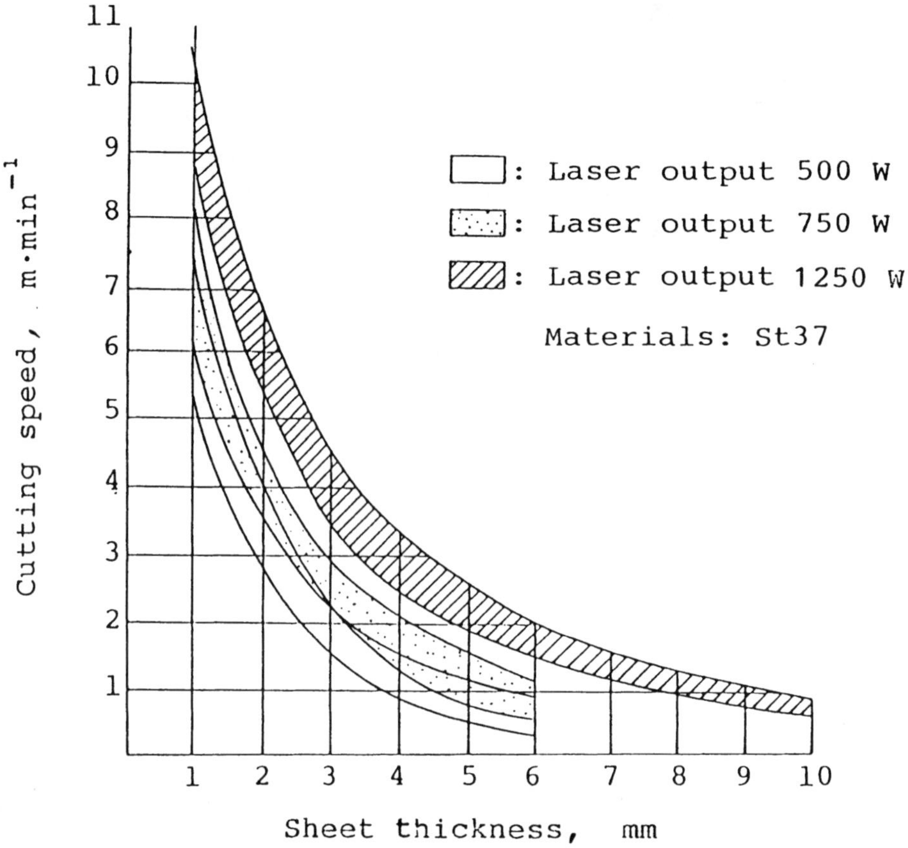

Fig. 4.4 Optimal Laser Cutting Speed for Various Sheet Widths[2)]

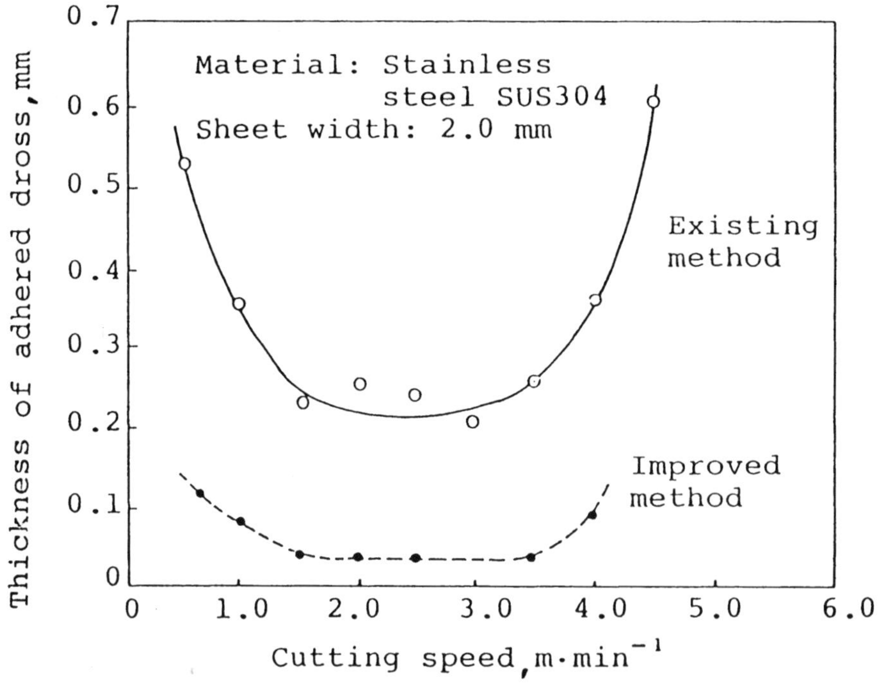

Fig. 4.5 Comparison of dross adhesion before and after (improved method) use of anti-sputter agent[3)]

of the aluminum alloy, the selection of an appropriate parameter can enable selective cutting of the plastic sheet only.[2] In the experiment done by the Metal Material Technology Laboratory, which was mentioned earlier, the reflectance can be controlled by surface treatment.

Recently, a successful instance of ceramics cutting was reported. The key to success is the selection of cutting conditions that cause an adiabatic field.

ZrO_2 shows favorable cutting results under all conditions, but Al_2O_3 is difficult to cut. This is said to be due to the differences in the temperature profile shown in Figure 4.6.[4]

Laser cutting of aluminum alloy, which was impossible due to the high reflection rate and thermal conductivity, entered the practical-use stage, offering the possibility of successful laser welding. This has attracted attention as the processing technology to permit aluminization of a variety of parts.

Komatsu Ltd. and Mitsubishi Electric Co. Ltd. can cut an aluminum alloy sheet as thick as 3 mm in a mode of single beam. The CO_2 laser processor used has a rated output of 1 kW. This permits cutting of a thickness of 9 mm of mild steel. Komatsu Ltd. uses continuous and pulse energy output; and can achieve optimal conditions by oscillator output control and adjustment of the component and flow quantity of the assistant gas. On the other hand, Mitsubishi Electric uses pulse output. A good cut surface is obtained by elevating the pulse peak output.

4.1.2 laser welding

Research on laser welding is being actively pursued. Especially, a large amount of research is being done on (a) higher output power for increasing laser-weld penetration, (b) use of a assistant gas or reduced pressure atmosphere for plasma removal, and (c) joint use of the TIG welding method.

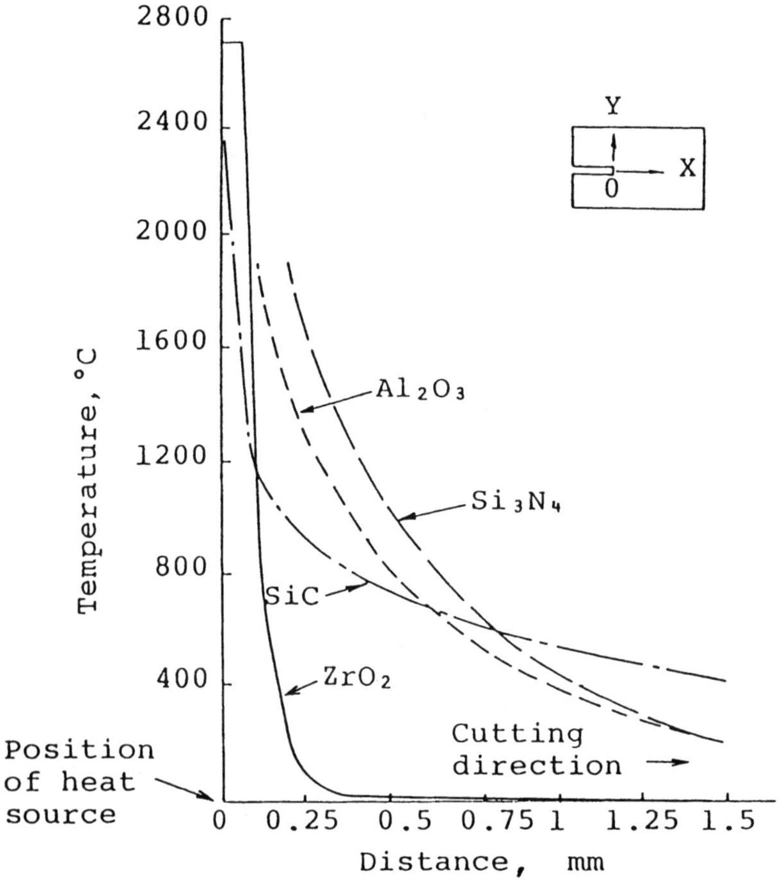

Fig. 4.6 Distance of temperature profile for several ceramics[4)]

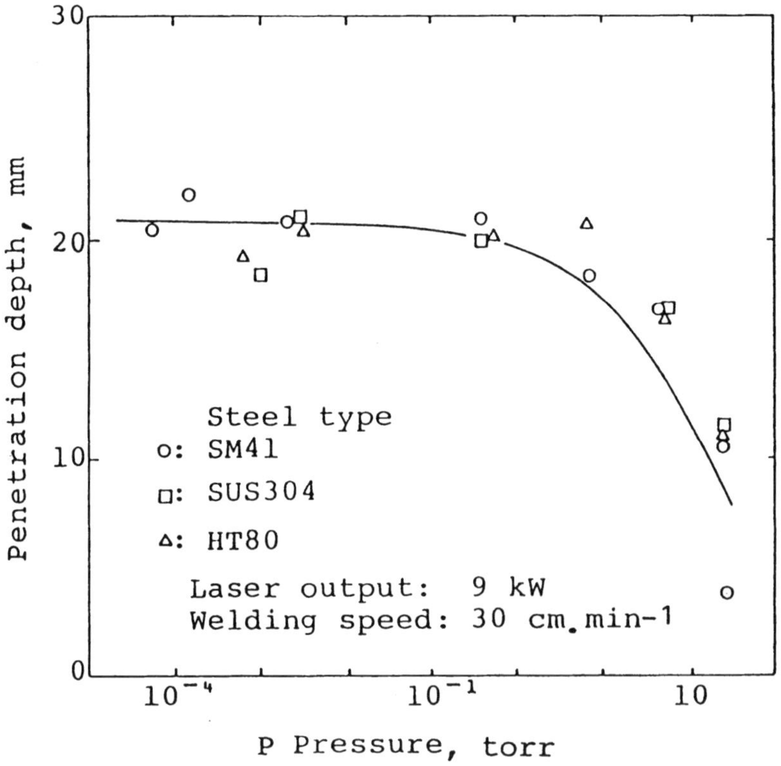

Fig. 4.7 Dependency of Weld Penetration Depth on Pressure in Laser Welding[6)]

Because lasers are of the same order as electron beams in energy density, laser welding is often compared to electron beam welding. But there is an essential difference. For electron beams, the absorption rate on the metal surface in a vacuum is 70%, whereas with CO_2 lasers, that absorption rate on the metal surface is 10% because laser beams are easily reflected.

Weld penetration by electron beams or lasers requires a critical energy density that melts the surface and produces a keyhole. With lasers, because reflection is strong on the solid metal surface and the absorption rate rises only after melting, the critical value is considerably higher than with electron beams.

Therefore apparently laser beams cannot compete with electron beams in weld penetration. Lasers, however, have features in precision welding and cutting, using the Nd-YAG laser which has a small focal point and short wavelength. But there is controversy in the academic world on the subject of comparative merits and demerits between laser and electron beam according to the object. In addition, by taking advantage of the optical nature of lasers, they are already used effectively in welding in locations impossible for workers to approach, for instance, for welding a steel belt coil on a pickling line. The high-speed welding of steel sheets, by high-power lasers on a mass production line is also an attractive theme to tackle.

The welding experiment conducted by the previously-mentioned Metal Material Technology Laboratory, shows that the fusion depth of welding performed under the conditions of 8 kW laser output power and 1 $m.min^{-1}$ welding speed is about 10 mm. The power conversion efficiency, the efficiency of beam energy versus depth of fusion, moreover, is 40 to 50% compared to that for electron beam welding. The bead width at the part welded by lasers in the above experiment is considerably wider than that produced by electron beams. This is probably due to (a) the attenuation of the beams by plasmas under atmospheric pressure, and (b) the weakening of clearance of melted metal according to the diminution of the metal vapor spouting output, the weakening of perforation thereby. Also, extensive blowholes were trapped in the fusion zone, especially at the root. Judging from the solidification layers at the cross section, this is probably a result of the fact that the gas flow at the center shield is pressed upon the melted metal at the root section, resulting in stagnating the melted metal and in an unstable flow.

In order to prevent the occurrence of such blowholes, the following countermeasures are possible: penetration welding, changes in shield gas flow, and agitation of welding metal by beam vibration.

Aluminum laser welding also generates pores. It was found that the gas within the pores is mostly hydrogen and that the primary hydrogen source is moisture in the atmosphere.

The laboratory of Fujitsu Ltd. is studying the lowering of welding crack susceptibility by making a joint site a eutectic structure with Si or Ni melted into the joint site by means of YAG laser, applying alloy filler or plating. There is a possibility of solving the welding crack problem and of applying the results of their study to the precise welding of electronic parts.

4.1.3 Experimental laser processing

a) Prevention of corrosion and improvement of the welded part of stainless steel

It is well known that the degradation (or weld decay) at a weld of 304 austenite stainless steel is difficult to trigger by electron beam welding. This is because electron beams enable single-pass welding of minute width and deep fusion. The effect is also expected with lasers when welding of thin sheets.

Elimination of weld decay requires solution heat treatment of the entire welded object. This is because local heat treatment by the existing thermal source regenerates weld decay adjacent to the weld. Lasers enable solution heat treatment solely at the surface layer without generation of weld decay.[5]

b) Carburizing treatment in controlled atmosphere

It is interesting to know how irradiation of laser beams under depressurized conditions or adjusting atmosphere affects metal processing. An increase of the degree of penetration due to pressure reduction is shown in Figure 4.7. It was found that surface welding by laser beams under depressurized conditions and with graphite applied to the steel sheet increases the carbon content remarkably (compared to that in the atmosphere) and raises hardness[6] , as shown in Figure 4.8.

c) Surfacing

Surface treatment by laser is the most interesting theme in processing technology. The tip of a prototype edge, manufactured by melting and surface cladding with powder containing high alloy and high carbon onto soft steel sheet is very effective in lumber processing. Dr. Masumoto and others reported that the result of apitong cutting test, in comparison with the tip of an SKH51 steel edge, indicated that as the quantity of vanadium carbide VC in the surface alloy increases, life expectancy of the tip of the edge becomes longer and very effective compared to existing edge tips.[7] [8]

d) Improvement of fatigue strength by weld bead reinforcement:

It is well known that compared with the fatigue strength of the parent flat metal plate, the fatigue strength of a welded joint is usually lowered by the notch effect due to some surface irregularity particularly at the end of the weld. Because lasers enable local surface welding or selective heat treatment within a very small area, an attempt is being made to use this feature to heat treat the welding termination point to improve fatigue strength.[9] Rewelding at the butt weld terminus of aluminum alloy (A5083P) at a speed of 500 or 800 $mm \cdot min^{-1}$ shows an improvement in fatigue life compared to the case when no such rewelding was done. But when treatment speed is low, cracks occur at the weld

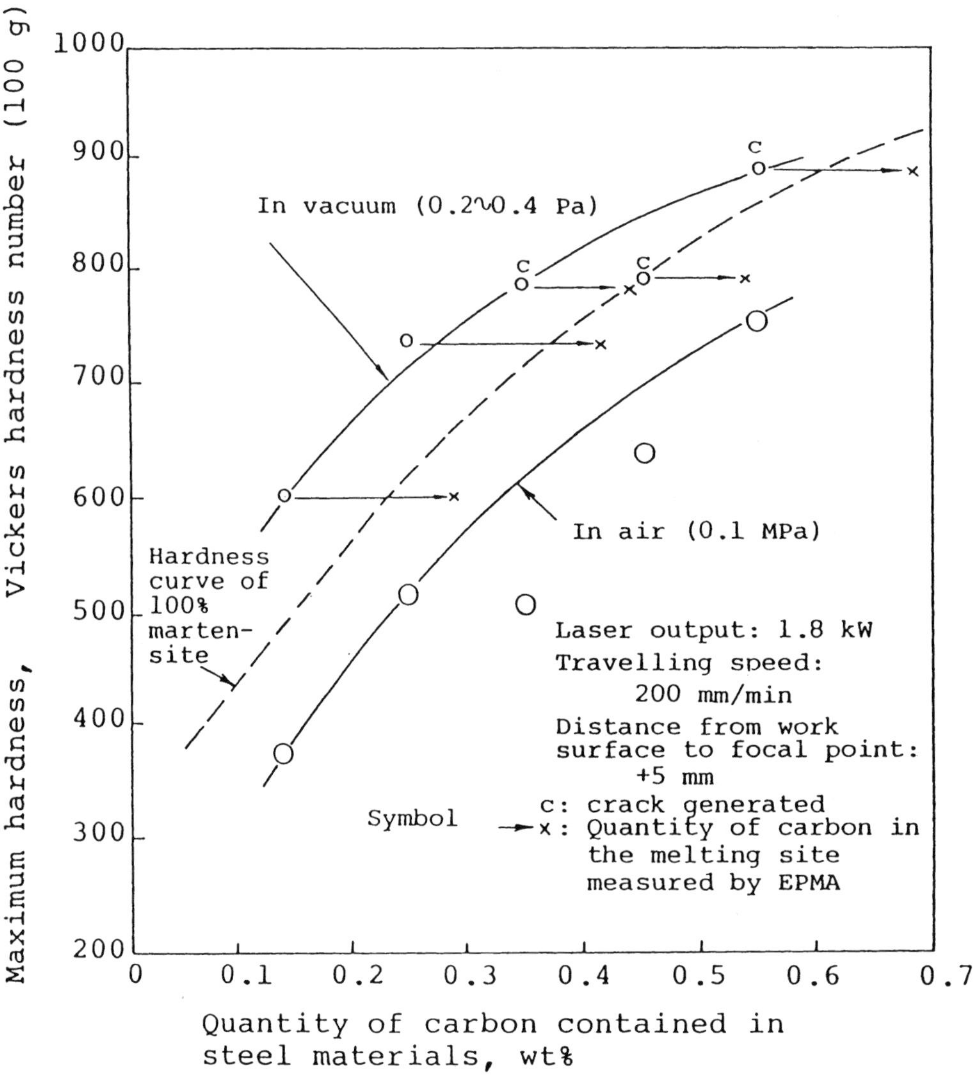

Fig. 4.8 Relationship Between Maximum Hardness and Carbon Quantity at the Laser-Melted Site[6]

toe, making little difference compared with the case where no such rewelding was done.

Lasers have established their reputation in cutting varied materials, and some practical examples are reported of lasers used for welding, surface treatment, and processing. At present, however, the potential of lasers in not being fully used in the processing technology field. Further information exchange and joint research will be required for full use of laser potential.

4.1.4 New experiments

a) Microtreatment

The light absorption peak of $Al(CH_3)_3$ or $Cd(CH_3)_2$ is in the neighborhood of 200 nm. These compounds undergo photolysis when exposed to ultraviolet light from an ArF laser. As a result, isolated Al or Cd atoms accumulate on the substrate to form a thin metal film. Metal films, SiO_2 films, Si films, or Ge films are formed through photolysis or photo-chemical reactions.

Doping atoms isolated by of the photolysis of doping gas by excimer lasers can be injected into the semiconductor substrate. Using this principle, attempts are being made to produce solar cells or change the electrical characteristics of substrates.

Ultraviolet excimer lasers are also drawing attention as light sources for lithography or holographic grains. Resolution of lithographic patterns by KrF or XeCl laser (1,000-line pairs nm^{-1}) is higher than that obtained by a conventional mercury lamp and can reduce exposure duration by about two orders of magnitude. It is also said that X-rays emanating from excimer-laser-generated plasmas are promising as a light source for X-ray lithography.

In addition, excimer lasers are being increasingly used as light sources for laser annealing, and recently manufacture of a prototype excimer laser annealing unit for solar cells has begun.

b) High-temperature superconducting thin films.

The Electrotechnical Laboratory (Director: K. Sugiura) of the Agency of Industrial Science and Technology has succeeded in the formation of thin films by fusing micropowders of a high-temperature superconductor with high-output laser beams.

A feature of this laser fusion technology is its ability to supply a very small quantity of materials (10-30 micrograms.sec^{-1}). This is supplied by a microfeeder (micropowder feed unit) developed by the Electrotechnical Laboratory

Material powder from the feed unit is fused by a 3 kW laser to form a film of 50-100 microns thickness on the substrate.

The film is not superconductive state as it is, but its electrical resistance disappeared at 80 K when it was heated for 12 hours at 840 ℃.

c) Generation of superconductivity by CO_2 laser

When irradiating a high focussed laser onto a gas specimen transparent to a laser with a wavelength, the light energy is almost completely absorbed and dielectric breakdown is generated. It also generates plasmas along with a keen sound and emission of high-brightness light. The group, whose chief research is Dr. K. Takeuchi of the Institute of Physical and Chemical Research, succeeded in creating super fine particulates from the vapor phase induced by the gas breakdown.

Specifically, mixed gas consisting of $BCl_3 + H_2$ is placed in an irradiation cell made of glass; then pulsed light from a CO_2 gas laser is focused by a BaF_2 lens. The threshold value of the breakdown fluence (laser energy per unit area) is represented by the fluence value at the focal point. This focal point is obtained from the relationship between the focal length of the lens and beam divergence after measurement of the pulse energy when the initial light emission from the target is recognized. Prior to light emission, the strength of the incident light is gradually increased by adjustment of the light depreciation plate in front of cell.

Dark brown particles created by the breakdown of mixed gas were found to be a spherical super fine particulates with a narrow particle distribution, of an average diameter of 80 nm. Also, absence of X-ray diffraction patterns and the creation of a halo pattern by electron diffraction indicated that the particles are amorphous boron.

Laser irradiation of a mixture of $Fe(CO)_2$ and SF_6 in the same process allows SF_6 plasma to decompose $Fe(CO)_5$ to create γ-ion particulates of average particle size of 0.1 μm. Although the structure of the particles is different from that of the boron particles, it was found that various particulates may be made with this process.

d) Clad steel tube

NKK (ex-Nippon Kokan K.K.) manufactured a prototype clad steel tube of high heat resistance and strength to which an corrosion-resistant alloy was externally molded. This is considered for use as the heat transfer tube of a heat exchanger in chemical plants exposed to strong corrosive liquids.

A corrosion resistant $Ni-C_r$ alloy powder such as Inconel 671 is discharged from the supply tube onto the outside of a rotating inner tube. The powder is melted by a CO_2 gas laser of maximum output of 5 kW. The inner tube is water-cooled to prevent material quality changes. The inner tube moves forward in the axial direction while being rotated, and its exterior is entirely covered by the spiral clad layer. Adjustment of the alloy powder supply speed, laser output, and rotational speed and speed of feeding the inner tube enables the thickness of the clad layer to vary within the range of 0.2 to 3 mm. The clad layer can be also made thicker by multilayering.

e) Research on utilization of the shock effect during laser irradiation

Light concentration of high output supershort pulses can raise the conversion efficiency of light energy to elastic energy more than 30%. The pressure of the generated shock wave is very high and is known to become about tera pascals. The value is a high transient stress that cannot be obtained by a static pressure device. Presently, basic research is being conducted, including the generation mechanism of shock waves by laser plasmas and on the pressure measurement method.

One active example of the use of the supershort pulse laser in material processing is the hardening of a solid surface. An attempt is under way to strengthen or harden Al alloy or stainless steel by laser-induced shock waves.

Impulse compression brings about a large temperature rise, compared to equi-entropy adiabatic compression. The state of impulse compression is released through adiabatic expansion by a tensile wave. The temperature at the time of restoration of the normal pressure state is residual temperature, not the initial temperature. When compression brings about a large-volume change, that is, when a substance of great compressibility or a substance showing phase changes accompanied by a large volume reduction is used as a specimen, a considerable temperature rise is expected.

A porous specimen or a specimen with uneven cracks indicates a large temperature rise at the time of compression. An extreme case may indicate expansion caused by shock wave pressurization. Conversely, use of these effects, that is, use of the high temperature generated on the interface of the particle or by filling the empty pores, will create a sintered ceramic body that cannot be obtained by static pressurization.

The use of temperature rise caused by impulse pressurization can cause crystallization of an amorphous alloy. The phase that appears in this case may be different from a phase obtained under normal pressure or static high pressure, or it may be a new quasi-stable phase. The problem is how to use the supershort pulse laser to generate supershort duration, superhigh pressure.

4.2 Creation of new materials and substances

The use of lasers to create new materials and substances was applied in laser nuclear fusion and isotope separation. Recently, lasers have also increasingly used in new fields such as materials synthesis or surface treatment and processing using laser-induced chemical reactions. The rapid extension of application fields is due to the rapid improvement of laser performance achieved during the past several years, such as the increase in the average output power brought about by the high repetitivity of excimer lasers and their prolonged life.

4.2.1 Film deposition by the light excitation process

Some interesting research examples in chemical reaction and application are given in this and the following sections.

(1) Amorphous silicon alloy film

An alloy film made of silicon and carbon or germanium is important as the material for tandem type solar cells, and control of the reaction is being attempted to manufacture high quality alloy films. According to Matsuda, the plasma CVD method with plasma control by the three-electrode structure and the dilution of material gas by hydrogen is effective for the manufacture of high quality alloy films.[10] Konagai reports that a mercury intensification light CVD using hydrogen-diluted raw gas also enables formation of good quality film.[11] It is further reported by Nishida and Suzuki that an excessive hydrogen dilution enables the formation of microcrystal films.[12],[13] Composition of an alloy film probably depends on the product of concentration of the active species in the vapor phase and the adsorption ratio. Tachibana estimates that the adhesion rate of CH_x radicals is about 25% of that of SiH_x.[14] Matsuda says that the surface reaction can be controlled by the selection of a hydrogen-radical concentration in the vapor phase or the vapor phase decomposition process.

(2) Sillicon epitaxial growth

The concept of light excitation epitaxial growth was proposed by Nishizawa in 1961.[15] After this, Tamasaki proved that epitaxial growth is possible at a temperature lower than in conventional method by reducing the epitaxial growth temperature below 600℃ This is enabled by the appropriate selection of the light source or material gas.[16] As the epitaxial growth temperature drops, however, the tendency to epitaxial-growth inhibition increases due to the influence of the natural oxide film or the adhered matter on the substrate surface, or the slowdown of surface diffusion.

Recently, Konagai and others have succeeded in lowering the temperature of the light excitation epitaxial growth substrate to 200℃.[17] Their method uses (a) silane fluoride (SiH_2F_2), (b) mercury intensification employing a mixture gas of a small amount of SiH_6 (or SiH_4) and H_2, and (c) uses a low pressure mercury lamp light CVD. It is reported that the use of fluoride material gas together with surface reaction control by hydrogen radicals produces on-the-spot etching of spontaneously generated oxide films.

(3) Film formation by photo-initiation

Formation of insulating film by photo CVD is being actively studied, principally for application into VLSI or compound semiconductor devices. Matsunami and others discovered the photo-initiation effect in the accumulation of SiO_2 film created by the reaction of monosilane and oxygen.[18] Ultraviolet rays are briefly irradiated at a substrate temperature that does not cause thermal reaction. Once the photo CVD has started, film accumulation continues without ray irradiation. Although discontinuing the material gas supply suspends accumulation, gas supply restarts it. This indicates, (it is reported) that an essential reaction site is formed by the initial irradiation and that it is preserved on the film surface. From the standpoint of catalyst chemistry, the activation energy considered to trigger Si-OH formation on the film surface is reported to be lower than the thermal CVD in a high temperature area.

The photo CVD for metal films uses patterned ultraviolet rays for (a) decomposition of the absorption layer of organic metal molecules, (b) selection and adhesion of the growth nucleus, and (c) ray irradiation over the entire surface or heating the substrate, thus permitting film growth only on the nuclear-forming site. Overseas also, some trials are reported of Al growth on Zn or Cd nuclei,[19] and Al growth on Al nuclei,[20),21] Also, Mikoshiba et al. found that ray irradiation of the substrate in parallel with the decomposition of trimethyl aluminum in the plasma vapor enables accumulation of carbon-free Al film only in the ray-irradiated portion.[22]

Light-induced catalytic reactions are also a developing field of study.

In general, when the initiation process requires maximal thermal energy and the succeeding growth process requires less energy, laser irradiation in the initiation process acts as the heating process required for growth reaction, thus enabling a lowering of the reaction temperature. This can, in particular, prevent secondary decomposition arising when the created substance is thermally unstable, achieving high selectivity and production yield. This also has the advantage that pulse-laser irradiation, which enables the selective excitation of only the reaction precursor, and the resulting energy saving, can create and recreate the catalyst activation species at quite short intervals of time, lending hope for obtaining highly active catalytic reaction under moderate conditions. Also, as with the chain reaction, turnover of the ray-created activated catalysts achieves a high quantum yield and permits reduction of the photon cost.

Ehrich[23] used ultraviolet ray radiation for the formation of pattern distribution of Ziegler catalyst reaction sites, and succeeded in the polymerization of ethylene and acetylene. The polymerization commences after (a) the scanning of ultraviolet beams onto $TiCl_4$ and $Al(CH_3)$ adsorbed on the substrate, (b) passing them into a vacuum, and (c) the introduction of ethylene as a reactant. Polymerization advancement is probably a result of the creation of catalytic activation species, caused by alkylation of titanium chloride under irradiation, and to repetition of the monomer coordination and insertion reaction.

Letokhov, a Soviet authority on laser chemistry, and other are studying the process of directly synthesizing amino acids from diazo-carboxylic acid and ammonia.[24] Irradiation by Nd: YAG laser (266 nm, 30 ps, 2 mJ) of a water solution of diammonia maleate produced aspartic acid at a quantum yield of 0.4. Laser irradiation of a fumaric acid salt or citric acid salt also produced of aspartic acid or methyl aspartic acid. The reaction is assumed to have taken place via the highly activated state resulting from dual-photon excitation. The maximum theoretical quantum yield being 0.5, 0.4 is a high value and indicates the possibility of practical application.

The Chemical Research Laboratory and Mitsui Toatsu Chemicals Inc. are now studying the laser processing in the synthesis of allyl chloride from 1,2-dichloropropane, a by-product of propylene oxide manufacture.[25] This is the dehydro-chlorination reaction similar to vinyl chloride's as cited in 3.3.1,(2). This laser process is a difficult to grasp because it has four types of reaction routes, and the allyl chloride to be created is the most unstable substance in the reaction. Thermal decomposition without laser irradiation requires temperature above 500℃, and the decomposition of the allyl chloride product chokes the reaction tube. KrF laser irradiation reduced the temperature to about 400 ℃ at the time of reaction commencement, producing allyl chloride of higher purity.

The same researchers studied hydrogenation by excitation of laser rays using an Rh- or Ir-complex compound containing triphosphine and found that irradiation by an excimer laser enables highly efficient hydrogenation of chain alkynes, unsaturated esters, and unsaturated alcohols.[26] The mechanism details are being studied and the reactions are expected to be developed as an example of the efficacy of laser high brightness and short pulses in catalysis.

IHI reported that laser irradiation at a wavelength of 350-450 nm of toluene and chlorine, mixed stoichiometrically, created alkyl-benzene containing no byproduct at a yield of 100% (Pat. unexam. 61-218536, 1986).

Furthermore, Eizai Chemicals and IHI laser irradiated vitamin K, which is a mixture of cis and trans isomers, and created trans isomers at a high yield. (Pat. unexam. 63-162649, 1988).

(4) Formation of ceramic films

Ceramic films are attracting attention as chemically-resistant protective films, transparent protection films for optical materials, and insulating films for semiconductor or electrical parts. So far, these films have been formed by vacuum deposition, sputtering or ion implantation. It has been recently reported, however, that photochemical reaction provoked by excimer laser has begun to be used for film formation. An example of this is given by Sugii and others who used irradiation by ArF laser of an Si wafer in an NH_3 atmosphere for the creation of Si_3N_4 film.[27] Overseas, XeCl laser irradiation of an Si wafer in an oxygen atmosphere was used to create SiO_2 films.[28] In another instance, ArF laser was used to photodecompose a gas mixture of SiN_4 and N_2O to form SiO_2 films on a substrate.[29]

A method of easily forming a transparent and tenacious SiO_2 film in air, which was used by Murahara, is explained here.[30]

As shown in Figure 4.9 (a), dimethyl siloxane silicon oil consists of Si and O combined in chain form. Two methyl groups (CH_3) are linked to Si. The silicon oil has a strong absorption band at 193 nm. The photon energy of the ArF laser is 147.2 kcal, the dissociation and binding energy wave of C-H is 81 kcal.mol $^{-1}$, that of O-Si 191 kcal.mol $^{-1}$, and that of Si-C 105 kcal.mol $^{-1}$. Therefore, irradiation by ArF laser beams dissociates Si-C and C-H, but does not dissociate O-Si. At the same time, the irradiation changes oxygen molecules in the air into oxygen atoms O in the ground state (3_p). These oxygen atoms instantly combine with Si from which the methyl groups are detached, to form the chain-shaped SiO_2 as shown in Figure 4.9 (b). A slide glass to which

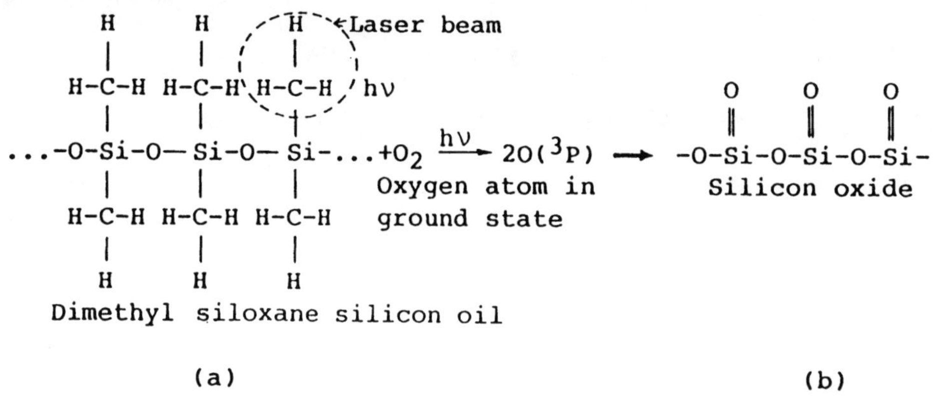

Fig. 4.9 Dimethyl Siloxane Silicon Oil and Molecular Structure After Laser Oxidation

silicon oil has been applied is put on the computer-controlled X-Y stage, then lens-focused ArF laser beams are guided onto the surface of the glass. The X-Y stage is thereupon moved by computer for character drawing. Wiping off silicon oil from an unexposed part after completion of the exposure formed a transparent, hard film 2μm thick that is impervious to cutting.

4.2.2 Laser annealing (high-speed crystal growth)

Fusion of the Si crystal surface by irradiation with laser beams of 10 ns pulse width causes an immediate high-speed crystal growth on the unfused substrate. A reduction of the pulse width to several ns causes the solidification speed to exceed the limit speed for crystal growth, creating an amorphous substance. Quenching of various stages in the solidification process by adjustment of growth speed will give us a strong clue to the mechanism of crystal growth. Also, on the applications front, quenching the transient process (from fused liquid to a solid) enables the extraction of new characteristics. For instance, Amorphous Si formed as described above is expected to have a greater density and a different nature from the ordinary substance.

Such crystal surface treatment by laser beams is known generically as laser annealing, The process was begun in 1974 in the U.S.S.R. to recover surface layers damaged by ion injection. The method has the following features when compared to heat treatment by electric furnace. These include the possibility of local heating. Impurities introduced by ion injection are 100% electrically activated even when they exceed the solution limit. For Si, open-air treatment is possible. These features enable simplification of semiconductor element manufacturing process and formation of a special structure by partial IC treatment, providing possibilities for the development of devices which cannot be obtained by heating the substrate as a whole.

For instance, an insulating film is put on an Si monocrystal subsrate, then a monocrystal Si film is formed on the film. Its application is being studied by Tamura and others, which will lead to the manufacturing of three-dimensional ICs.[31] This technology is called Silicon on Insulators (SOI).

4.2.3. Photo-etching

Horiike et al. reported etching polycrystal Si by Cl_2 with an XeCl excimer laser.[32] In this case, the n-type is etched isotropically and the nonadditive-type and the p-type are etched anisotropically, and the etching velocity diminishes in that order.

For the non-additive or p-type, electron from Si moves to chlorine radical generated on the surface of Si by beam irradiation, resulting in reaction between chlorine ion (Cl^-) and Si^+ to progress etching. Therefore (a) only that portion of surface is etched on which beam is irradiated, and (b) anisotropical etching is formed on the side wall because no beam is irradiated on it.

For the n-type, since its electron concentration is high even with no beams irradiated, Cl-ions are created on the surface and diffused in bulk. Etching then becomes isotropical.

Hirose at al. performed ArF excimer etching for Si[33] and SiO_2[34] using NF_3 and found etching behavior different from that for the Cl_2 mentioned above. The group is now studying the mechanism by surface observation based on the XPS analysis, where Si etching is performed with F radicals. Based on the attribution of extension of the Si_{2p} peak to the fluoride substitution number having a great electronegativity, and on the dependency of the peak strength on the etching duration, the following mechanism was proposed. The fluorine radical is sufficiently small to penetrate between the Si lattice and exhibits a greater electronegativity than Si. This attracts Si valency and forms an Si-F bond of high ionic bond strength. With the passage of time, the number of F radicals combined with Si increases, to disappear into the vapor phase in the form of volatile molecules of SiF_2 or SiF_4. Unlike Cl radicals, since F radicals are ionized by valence electron movement, isotropical etching is produced on both p and n type areas.

4.2.4 Materials extraction, refining, and synthesis

The refinement method, similar to the isotope separation method, uses selective laser irradiation of impurities to convert them into reactive substances that facilitates easy elimination of impurities. In this method, the system need not be heated, vaporized, or separated by distillation or membrane, and chemical products to be refined need not consume energy and may be refined without alteration or loss. The laser method only requires an energy supply sufficient to cause reaction of only a very small quantity of impurities and is ecomonically efficient. Also, this method has a separation coefficient sufficiently high to require only one stage, thus being especially liable to produce ultra-high purity. Research is being conducted on the purification of gas used for ultra-high purity electronic materials, or elimination of dioxin, which is deadly poisonous substance and a secondary product created during synthesis reactions (Table 4.4).

An interesting example of laser treatment is the irradiation of micro-organisms by ultraviolet laser for the selective extraction of ferments, amino acid, or proteins. This was done by Ishikawajima-Harima Heavy Industries and the Nippon Zeon Co. The process is expected to be applied to more materials as a chemical treatment for thermally unstable organism substances (Pat. unexam. 62-51990, 1987).

With the background mentioned above taken into consideration, an IHI tunable laser has been developed as an experimentation tool to facilitate the search for an optimal wavelength that constituted a bottleneck in the basic

Table 4.4 Purification of Chemical Products by Laser

Impurities	Main compoment	Laser used
CH_2ClCH_2Cl, CCl_4	$AsCl_3$	CO_2
$COCl_2$	BCl_3	CO_2
B_2H_6	SiH_4	CO_2
DF_3Cl, CF_2Cl_2	CF_4	CO_2
CO_2	N_2+He	CO_2
PH_3, AsH_3	SiH_4	Excimer
$COCl_2$	BCl_3	Excimer
H_2S	$CO+H_2$	Excimer
Dioxin	Cholorophenol-Derivatives	Excimer
Eu	Lanthanide group	N_2, Ar^+

research on laser chemical reaction (Pat. unexam. 60-172347, 1985 and 61-218536, 1986).

The IHI tunable laser has the following features. The wavelength is variable within a wide range, from the ultraviolet (196 nm) to the infrared (4000 nm). The maximum peak output at 196 nm is 300 kW. Simultaneous oscillation at two wavelength is possible.

IHI announced completion of the tunable laser in 1983. A large number of companies interested in laser chemistry then submitted a host of enquiries. Basic tests were jointly conducted on some of the applications. The breakdown by types of reaction is as shown below.

	number of test items
General-purpose chemical products	5
Fine chemicals	4
Polymers (surface treatment)	1
Removal of impurities	2
Surface reaction	3
Extraction of biosubstances	5
Application to medical and pharmaceutical products	3

According to Katayama, most of the tests were conducted jointly by IHI and the inquiring firms, although some tests were executed by IHI alone.

It is reported that in the absence of consideration of the feasibility of manufacturing on a commercial basis, the test results were satisfactory for the most part and many of them led to applications for patents.[35]

Silane (SiH_4) gas, which is becoming popular for integrated circuits or solar cells, contains impurities such as arsine (AsH_4) and phosphine (PH_3).

Irradiation of silane gas by ArF laser causes a photo-chemical reaction which transforms the impurities into a higher molecular compounds, thus enabling their separation from the SiH_4 gas. The method can reduce concentration of impurities from 50 ppm to 0.5 ppm.[36]

An attempt is also being made to use Raman shift radiation of ArF or KrF lasers for photodecomposition of H_2S impurity contained in mixed gases used for synthetic chemistry (H_2 + CO). This produces high-purity refined gas.[37) Isotopes other than uranium are important for nuclear reactor materials (Ti, Zr, Mo, B, etc.), radiation sources (Cs, Cm, etc.), and tracers (I, Si, S, C, N, etc.) for superhigh-value-added chemical products. An important process is the enrichment of silicon 28, 29, and 30 isotopes by irradiation with a CO_2 gas laser of the working material Si_2F_6, carried out by the Institute of Physical and Chemical Research and the Shin-Etsu Chemical Co. Ltd. Arai and others report that laser beams oscillated by a hybrid gas of CO_2 and He very efficiently cause the working material to form SiF_4, which can be easily separated. This method is consequently already at the practical-use stage.[38)]

At the present time, isotope separation for most chemical elements of large atomic numbers can only be made on an extremely small scale by mass spectrograph. Accordingly, the possibilities for laser treatment are bright. Also, the absence of other competitive processes makes its applicability potentially high. The problem is to find a working material that corresponds to the effective wavelength of the laser and has a high separation coefficient.

Rare metals, especially rare earths, are attracting more and more attention as an additive component for future advanced materials. Refinement, however, by chemical separation is difficult. The Japan Research and Development Center for Metals opened an investigative sectional meeting in December 1987 to discuss whether or not research on rare metal related technology, including "high purity refinement of rare metals by laser excitation", should be conducted as a national project.

4.2.5 Metal alloys

T.F. Deutsch and others at MIT propose a manufacturing method for film-shaped solar cells by doping B onto the Si surface with irradiation by ArF laser beams of Si in a BCl_3 atmosphere. Murahara and others used the same method to examine the possibility of making an alloy by doping with a large amount of B.[39]

BCl_3 strongly absorbs ArF laser irradiation at 193 nm. Photon energy at this wavelength is 147 kcal. Furthermore, B-Cl association and dissociation energy is 131 $kcal.mol^{-1}$. The binding energy is smaller than the photon energy and is in the absorption band, making possible the easy extrication of B atoms by ArF laser beams.

ArF laser beams are focused onto the surface of the Si wafer placed in the BCl_3 gas atmosphere by using a lens. In this way a great amount of B is freed at a focused site having a high energy concentration. On the other hand, the beams of this wavelength penetrate Si only by 50 Å and the surface absorbs them at an extremely shallow depth to generate heat. Into this area B freed by photodecomposition is diffused, making an alloy. Repetition of the operation enables deep doping. It was found that selection of the laser beam energy and BCl_3 gas pressure can control the B content and its diffusion depth.[40]

4.2.6 Summary

Example of material processing were given and explained, mainly in terms of photo sensitivity or electronic excitation by ultraviolet rays. It is hoped that understanding and control of the reaction at the molecular level, such as consecutive synthesis of the laminated film by a combination of low-temperature epitaxial growth and plasmas, will lead to new developments in the beam excitation processes as shown in Table 4.5.

Table 4.5 New Processing Technology Using Lasers

Processing Type	Processing method	Laser used	Examples of processed objects
Vaporization	Laser PVD	CO_2, YAG	Ceramics, B, HgCdTe
	Hardening by impulse	CO_2, YAG, Q. glass	Al alloy
	Forming	CO_2, YAG	Si_3N_4, SiALON, Al_2O_3, quartz glass
	Cleaning	Q. ruby	C, O (on Si)
	Removal of burrs	CO_2, YAG	FRP mold, gear
	Etching	CO_2, YAG	Etching and Engraving for plastic and paper
	Ink drying	CO_2	Printing paper
	Rheology processing	CO_2	Viscous powder (metal, ceramic, plastic)
Melting	Synthesis of new materials	CO_2	ZrO_2, Y_2O_3, HfO_2
	Single crystallization	CO_2, YAG	Al_2O_3, ZrO_2, LaB_6, Si
	Fiber growth	CO_2	Al_2O_3, YAG, $LiHbO_3$, glass
	Recrystallization	Ar	si
	Annealing	CO_2, YAG, Ar, Excimer	Si. GaAs, magnetic garnet film, CdSe, Al_2O_3
	Polishing	CO_2	Molten quartz
	Melting and hardening	CO_2	Cast iron, tool steel
	Alloying	CO_2, Ruby, YAG	Metal/metal
	Mixing	Ruby	Ni/SiC
	Cladding	CO_2	Stellite, stainless steel, high-speed steel/carbon steel
	Glazing	CO_2, YAG, Ruby	Pd-Cu-Si, Fe-Ni-P-B, Ni-Nb, Al
	Joining	CO_2, YAG	TiN/Ti, ceramics/Al, steel
	Ceramic spraying	CO_2	ZrO_2, TiO_2, TiN
	Ceramic welding	CO_2	Al_2O_3, mullite, forsterite
	Melting and cutting	CO_2	Steel sheet

(Continued)

Processing Type	Processing method	Laser used	Examples of processed objects
Change in solid-phase	Strain relaxation	CQ_2, YAG	Steel, Al
	Strain control	CO_2, YAG,	Plane lens (acryl-polymer), microlens (glass), embossing design (stainless steel), plastic working (steel)
	Defect control	CO_2, YAG	Si (pn junction formation)
Chemical reaction	Fine powder fabrication	CO_2, YAG	Si_3N_4, SiC, Si, TiN, TiO_2, metal
	Laser CVD	CO_2, Excimer, Ar	Al_2O_3, TiC, TiN, SiO_2, Al, Si, GaP
	Plating	Ar, Kr	Cu, Ni, Au, Pd, W
	Resin hardening	Excimer	Platemaking, dies, dental filling agent
	Light sensitiveness	Ar	Relief printing, offset printing
	Etching	YAG, Ar, Excimer, CO_2	Al_2O_3, TiC, SiO_2, garnet, GaAs, Mo, W, Si
	Ignition	YAG	Propane, methane

4.3 Laser generator

Thirty years have passed since the invention of the laser (Light Amplification by Stimulated Emission of Radiation) by Maiman, an American, by extension of maser concepts into light. At present, four types of solid lasers are available: ruby, YAG, glass, and alexandrite. Other types include gas lasers such as CO_2, CO, Ar lasers, the excimer laser, copper vapor laser, and the dye laser.

An outline of the performance of these lasers was given above in Table 4.1. Presently, only CO_2 and YAG lasers can be used for production from the viewpoints of durability and economic efficiency. The YAG laser is a solid state laser. It is more compact than the CO_2 laser. It has an oscillation wavelength of 1.06 μm (near infrared), and is available in both continuous and pulse modes. Since the wavelength is near the optimal transmissivity for optical fibers, YAG laser beams can be transmitted by optical fibers, giving them an advantage in three-dimentional processing.

The power conversion efficiency, however, is about 3% and problems like cooling make it difficult to secure a large output from a single rod. An attempt is being made to solve these difficulties by using the gadolinium-scandium-aluminium-garnet, crystal with Nd^{+} and Cr^{+} added for improvement of the power conversion efficiency.

Another possible solution lies with glass material with Nd^{+} added and of a thin ladder-shaped that can give large surface for heat emission. Its thermal conductivity is still bad, however, and its cooling and highly-repetitive oscillation are difficult to implement. A high power slab-type laser is being developed that is said to solve the foregoing problems.

On the other hand, the CO_2 laser has a 10.6μm oscillation wavelength which is a far infrared beam with about 10 times wavelength of the YAG laser, enabling 20 kW class continuous output, which is huge. In addition, it features extremely high power conversion efficiency (more than 15%) compared to the other gas lasers, solid state lasers.

MITI's National project, "Flexible Manufacturing System Complex Provided with Laser", developed a TEA CO_2 laser, and the companies participating in the project made efforts to achieve high output power. The market, however, has showed a high demand for 0.5-1.0 kW medium-output devices. To accommodate this demand, action is being taken by marketing a triple-axis silent discharge system (Mitsubishi Electric Corp., Komatsu Ltd.), or a dual-axis system using the spare discharge system (Hitachi Ltd., Fujitsu Ltd.). These developments were not taken up as research objectives by the project.

Regarding subsequent high-capacity device development, the Industrial Research Institute (Chairman of the Board of Directors: Mr. Gaishi Hiraiwa) has announced the result of cutting experiments on metals, and other materials by means of an industrial CO (carbon monoxide) laser processor developed in cooperation with Mitsubishi Heavy Industries Ltd. The CO laser is next-generation-type laser that will replace the popular CO_2 gas laser. In this field, only Ishikawajima-Harima Heavy Industries has been developing oscillator, and its experiment in practical application conducted in combination with the machining processor is a world first. Reports indicate that the experiment has shown that the CO laser can be used as a good processor for metals and other materials.

According to Dr. K. Fujioka, ex-Keio University professor, who is responsible for development at IRI, both high output (maximum output: 1.3 kW) and high efficiency (electric conversion efficiency 27% [vs. 15% for CO_2 gas laser]) were achieved by the improvement of the discharge electrodes and laser resonators, as well as by speeding up gas-flow. Based on this, cutting tests were conducted on different metals. The results showed a high cutting ability, which for carbon steel is three to five times that of the current CO_2 gas laser. For stainless steel, the improvement is approximately 200%.

This means that when metal cutting is made with the same output set for both lasers, and the same object is to be processed, the CO laser can process faster than the CO_2 gas laser. Furthermore, when the cutting is done within the same period of time, the CO laser can cut thicker metal.

The cutting ability greatly depends on the wavelength of the laser. The wavelength of the CO_2 laser is 10.6 μm. That of the CO laser is 5.3 μm, which reduces the divergence angle of the laser beam to half that for the CO_2 gas laser. This permits the use of smaller optical parts for beam transmission and reduction of the focusing-spot diameter to half that of the CO_2 device. Only 1 kW is required by the CO laser for cutting, as compared with the 3-5 kW needed by the CO_2 gas laser. Also, the absorption rate of laser beams on the metal surface is higher at the 5.3 μm wavelength, providing excellent processing ability.

In cooperation with Mitsubishi Heavy Industries, the Industrial Research Institute has completed a 5 kW CO laser processor that is larger than the previously mentioned CO laser processor. The oscillation test has also been completed. The cutting experiment using a large machine was scheduled to begin within the 1989 fiscal year. The Laboratory will continue to analyse the CO laser's characteristics in future, but Mitsubish Heavy Industries is probably the leading contender for future merchandising. 20 kW CO lasers will be supplied in June 1990, after which the devices will be subjected to experiments for cutting stainless steel sheets of 300 mm thick.

The Institute of Research and Innovation's CO development is related to the nuclear reactor pressure container cutting technology verification test that MITI entrusted to, and which is being conducted by, the Atomic Energy Engineering Test Center. The development project has been applied as a technological measure to facilitate the dismantling of nuclear reactor. A lining is being made for nuclear reactor pressure container by using 400 mm low-alloy steel and 10 mm stainless steel. The cutting verification test requires a CO laser.

The Mechatro Comprehensive Development Dept. of Yokohama's Technology Laboratory, a part of the Ishikawajima-Harima Heavy Industries Co. Ltd., achieved a maximum power conversion efficiency of 25.1% at 3.1 kW output by their CO laser. Continuous oscillation of more than one hour has also been achieved, accelerating development towards practical use. The problems requiring solution are said to be improvement of the cooling system up to -120℃, high-speed cutting under fixed power, and cost reduction.

5. Market Trends

5.1 Prospects for the machine tool industry

Economic friction is intensifying as a result of the international trade imbalance. And, as the yen becomes stronger, Japan's economic structure shows signs of reform. Under such circumstances, a report by the economic structural adjustment study society for international cooperation (the Maekawa Report), and also a report of the basic plan for the industrial society of the twenty-first century stated that the most pressing problem is the realization of a cooperative international economic structure. The reports suggested the necessity of maintaining the vitality of Japan's industries to help solve the problem.

Regarding the future prospects of the processing machine market, which is one of the leading laser technology application fields, the round-table conference on future prospects of the machine information industry, held under the auspices of the director of the Machine Information Industry Bureau of MITI, issued the below interim report following continuous study that began in October 1986 (see Table 5.1).

5.1.1 Demand trends

a) International demand trends

Table 5.1 Committee Members of the Round-Table conference on Future Prospects of the Machinery Information Industy

Mr. S. Akazawa	Chairman of the Board of Directors of the Japan External Trade Organization(JETRO)
Mr. S. Asai	Chairman of the Board of Directors of the Japan Machine Tool Builders's Association
Mr. T. Ishii	Professor, University of Tokyo
Mr. H. Inaba	Chairman of the Board of Directors of the Industrial Laboratory
Mr. J. Enjoji	Advisor to Nihon Keizai Shimbun Inc.
Mr. R. Kaku	Chairman of the Board of Directors of the Japan Business Machine Makers Association
Mr. Y. Kosai	Chairman of the Board of Directors of the Japan Economic Research Center (JERC)
Mr. M. Kono	Vice-Chairman of the Editorial Committee of K.K. Yomiuri Shimbunsha
Mr. S. Saba	Counsellor to Toshiba Corp. Ex-Chairman of the Board of Directors of the Electronic Industories Association of Japan
Mr. S. Toyoda	Chairman of the Board of Directors of the Japan Automobile Manufacturers Association Inc. (JAMA)
Mr. N. Hashimoto	Chairman of the Board of Directors of the Communications Industry Association of Japan (CIA-J)
Mr. A. Morita	Vice-Chairman of the Board of Directors of the Federation of Economic Organizations (Keidanren) Chairman of the Board of Directors of the International Industrial Cooperation Committee
Mr. M. Warashina	Vice-Chairman of the Board of Directors of the All-Japan Private Sector Labor Union Council

A worldwide demand forecast is difficult to compile, but it is assumed, based upon past trends, that a slow increase is to be expected in the medium- and long-term outlook. On the other hand, predictions are that an increase will be seen in the supply by NIEs (Newly Industrializing Economics) of non-NC machine tools and general-purpose NC machine tools. As a result, a partial oversupply may occur.

b) Domestic demand trends

Domestic demand was stagnant for some time. Emergency economic measures, however, have brought about an increase in investments in manufacturing industries, such as general machinery, automobiles, electrical equipment, and precision industries. The recovery of demand has centered upon domestic sales. According to a survey by the Japan Machine Tool Builder's Association, the amount of orders received for machine tools in the January to June period of 1988 ¥462.9 billion(£1.78 billion or $2.89 billion), which represents a 45% increase compared with the same period of the previous year. A detailed breakdown indicates a 28% increase in exports and a 62% increase in domestic use. This provides graphic evidence of a completely domestically-oriented economy. In addition, considering the recent rising tendency, the annual amount of orders received has strong probability of attaining a total of ¥900 billion(£3.46 billion or $5.63 billion), and this category is expected to achieve the second highest level in history after 1985.

c) Japan's trade trends

It is assumed that to avoid trade friction, only a very slow increase will be seen in exports. As for imports, predictions are that they will remain at the same level or increase only marginally for the time being. This is a consequence of a cooling-off of domestic plant and equipment investment. In the future, however, imports are expected to increase at a rate higher than that for the upturn in domestic demand.

5.1.2 Technology trends

a) Automation

In the future, advancement is expected in the automation and systematization technologies for factory automation (FA), along with the adoption of network technology and artificial intelligence.

b) Higher precision

High-technology industries require extremely high processing precision. Attainment of this precision will require advancement in the development of high-rigidity design and contactless, machine-mounted measuring technology.

c) Action promoting new materials and processing technology

Machine tool technology will advance as new materials come into greater use. This advancement will include expansion of the cutting processing area by improvement of machine rigidity, plus tool development. Diversfication and advancement of processing technology using lasers and ions will also progress.

Five hundred and seven machine tools were exhibited in October 1988 by 96 member companies of the Japan Machine Tool Builder's Association at the 14th Japan International Machine Tool Fair. NC machine tool applications occupied 86% of the exhibits, but special processing machines such as discharge or laser processing machines also made an appearance, constituting 10% of the presentations.

No large-scale Flexible Manufacturing System was displayed as a production system. This indicates that this field is already in the mature stage and that units are being tailored to the needs of individual users.

5.2 Trend in the laser processing machine market

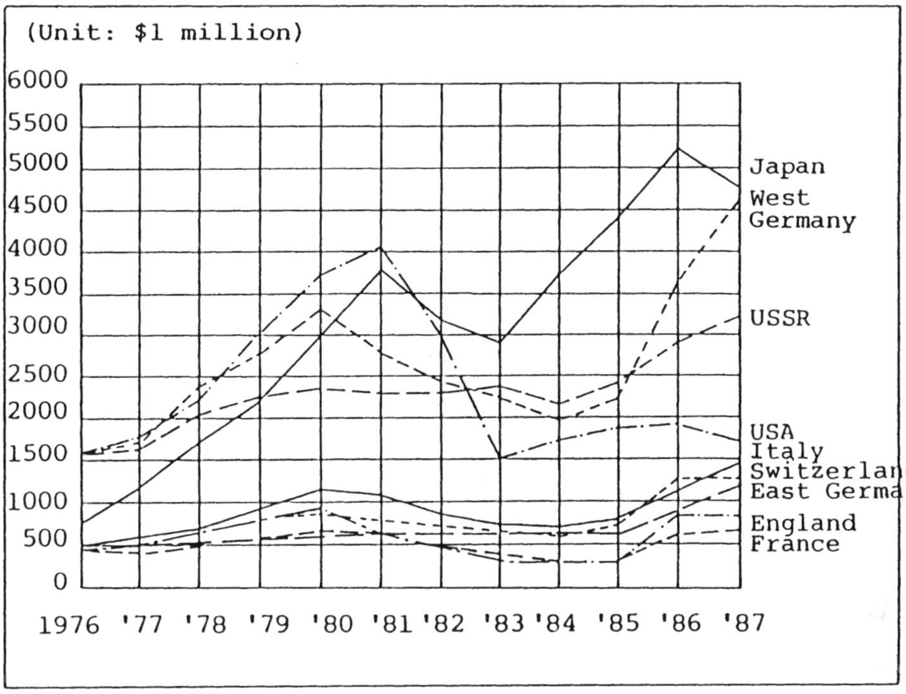

Provided by the "American Machinist"

Fig. 5.1 Production Trends for Cutting Machine Tools in Major Countries

CO_2 and YAG laser processing machines occupy most of the market. External factors such as the semiconductor depression and the strong yen brought about reductions in plant and equipment investment by makers of semiconductors, automobiles, and electric equipment, applying a brake on market growth. The number of orders received, however, has been expanding since 1987 (see Fig. 5.2).

Of the CO_2 laser processing machines shipped in fiscal 1986, the application fields are as follows.

Application field	Share
Metal cutting, drilling	80.7%
Nonmetal cutting, drilling	8.7%
Welding	6.7%
Heat treatment	0.9%
Research and development	3.0%
	100.0%

The composition by demand field is as follows.

Demand field	Share
Machinery	23.6%
Job shops	21.4%
Electrical equipment	21.0%
Automobiles	12.3%
Precision equipment	9.6%
Non-automotive transportation equipment	2.3%
other	9.8%
	100.0%

Recent product development trends are as follows:

1) Miniaturizing oscillators and systems

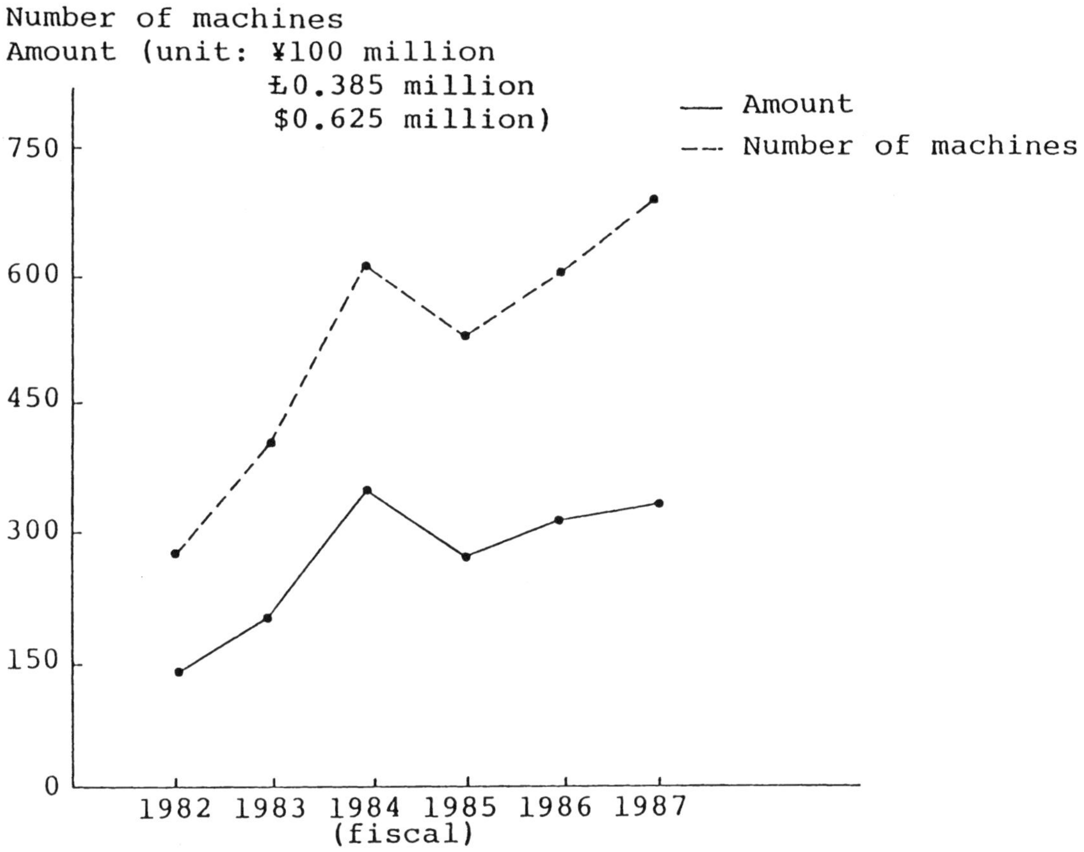

Fig. 5.3 Changes in the number and value of shipped CO_2 laser processing machines in Japan

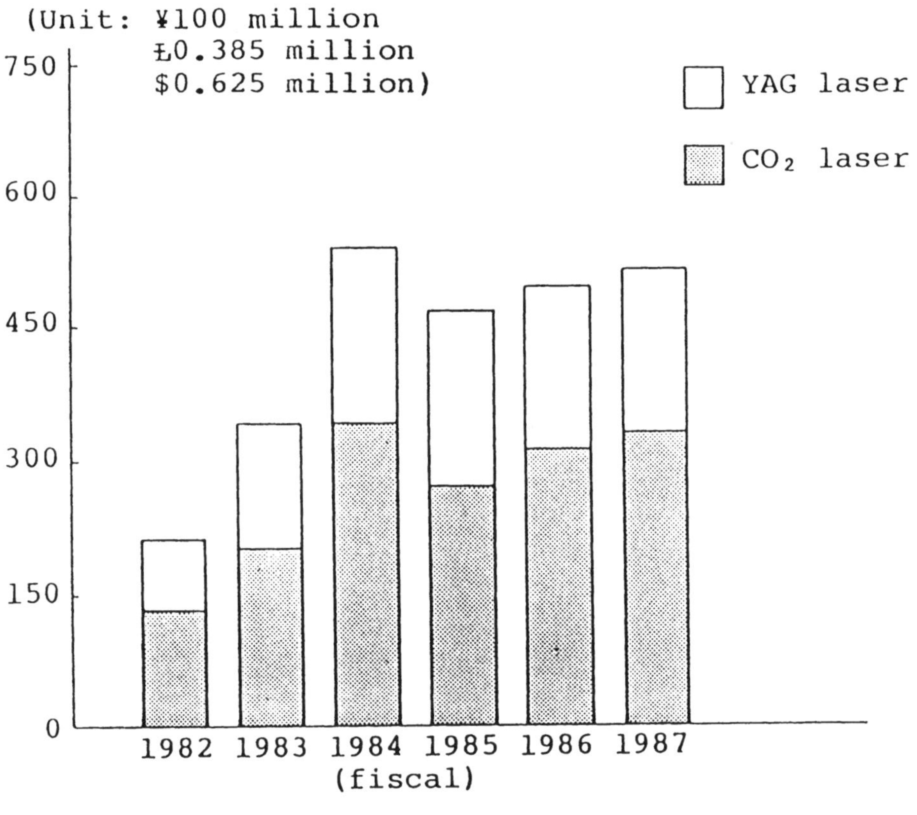

Fig. 5.2 Changes in the Japanese market for laser processing machines

2) Higher speed and precision

3) Cutting machines for non-ferrous materials such as aluminum and copper; also as stainless steel oxygen-free cutting machines

4) Operational improvement by combination with CAD/CAM

5) Development of three-dimensional processing machines

More than 50 makers submitted entries in this field. Even after 1985, ten new makers entered. As shown in Figure 5.3, such an intensification of competition produces a deviation between the number of shipped laser processing machines and the amount of machines produced.

In fiscal 1987, the Mitsubishi Electric Corp. enjoyed the top share, which represents about 20% of the market (145 machines, ¥6.5 billion (£25.0 million or $40.6 million). Mitsubishi manufactures a variety of products ranging from oscillators to NC devices to work tables, providing optimal systems tailored to customer's orders.

Oscillator output, furthermore, covers 500 W-10 KW in seven stages, aimed at covering a variety of industries and applications such as cutting machines, three-dimensional processing machines, welding machines, and surface treatment machines. Amada Co. Ltd., the leading maker of steel metal-working machines, holds the second highest share which is about 14%, showing a favorable trend with emphasis placed on high-precision processing machines and three-dimensional processing machines. The Shibuya Kogyo Co. Ltd., the largest maker of bottling plants, holds the third highest share (7%) and is active in the multilateral development of CO_2 lasers, YAG lasers, TEA-CO_2 lasers and others. The company is overwhelmingly strong in the TEA-CO_2 laser marking system market. Toshiba Corp. and Komatsu Ltd. enjoy shares of 6% each.

By categories, 32% of the YAG laser processing machines are used for trimming, 23% are employed for marking, 18% for welding, 10% for soldering, and 6% for cutting and drilling.

The YAG laser processing machines are mostly used by the electronics industry, including semiconductor makers. The machines are used in IC trimming, mask and memory repairing systems, and silicon wafer marking systems. The scanning-type marking system is mainly made for automobiles and the mask-type marking system is shipped to electronic parts makers for plastic work.

In the welding field, processing machines are shipped to electrical equipment makers for welding electron guns for cathode-ray tubes, printer's printing heads, relay parts, and others.

About 30 companies have entered the processing machine manufacturing market. Ten or more new firms have joined the competition in the past few years. The top share (51%) is owned by NEC Corp., which shipped 450 machines in fiscal 1987. The company's strong point is, as with Mitsubishi's mastery of the CO_2 laser, that NEC can provide a variety of products ranging from oscillators to systems to peripheral equipment, and the company is backed by a powerful sales organization and influential agents.

Toshiba comes in second with an 18% share and Fuji Electric Co. Ltd. is a distant third with 7%. Regarding the scanning-type laser-marker market, however, Fuji Electric enjoys a share of more than 90% and a near monopoly of the market.

The recent trend in technological development of Toshiba's YAG laser equipment is for achieving higher oscillator output. It is reported that the development in 1984 of a 600 W pulse laser oscillator and a 600 W continuous laser oscillator, and the development in 1986 of a 1.4 kW oscillator with four 400 W laser heads arranged in one resonator, enable applications for quenching, surface modification, and other operations, in addition to cutting and fusing. It is expected that this will lead to an increase demand.

6. Conclusion

An outline has been given of the organizations, dealing with Japan's laser technology related mainly to materials development. Details of development have also been provided.

Because inexpensive NC machines were developed and mass-produced, machine tools made in Japan, excluding laser-based devices, achieved rapid sales expansion beginning in 1980. Exports were mostly to the United States and Europe, and Japan became No.1 in sales worldwide by forging ahead of the United States and West Germany. In the light of the intensified trade friction, however, America's National Machine Tool Builder's Association (NMTBA) instituted an action in May 1985, claiming that increased imports of Japan-made machine tools will cause grave problems in terms of national defense. Since January 1988, therefore, Japan has been exercising self-restraint on the export of NC lathes and machining centers (MCs) to the United States.

In support of the National Military Establishment, the NMTBA set up the National Center of Manufacturing Science (NCMS) at the end of 1987 to push technological development in this field.

To cope with this development, the Japan Machine Tool Builder's Association (110 member companies; Chairman of the Board of Directors: S. Asai, Chairman of the Toyoda Machine Works Ltd.) reported on October 11, 1988 that it had commenced joint studies on the basic technology toward development of next-generation machine tools. The research themes are (a) development of supermultifunction machine tools that can overcome the restrictions that the present NC units impose on machining processes, and (b) development of more reliable machine tools that employ new materials. As a first step, prototype machine tools will be made. This effort will require three years of cooperative research by makers and universities. The laser technology developed up to this point will certainly be used in this quest.

Such government support will enable technological development that is too great a financial risk for a single company to undertake, and will trigger expansion of Japan's technological base by cooperative development involving many powerful companies. Each company will be able to continue to produce good results after completion of the project by further, independent development.

The fundamental research on laser technology is being conducted by the Institute of Physical and Chemical Research, while backup studies associated with the processing equipment are being executed mainly by Osaka University.

Leading processing equipment makers (Mitsubishi Electric Corp., Toshiba Corp., NEC Corp., Fuji Electric Co. Ltd., and others) already independently produce sufficient oscillators for present requirements. The problem, therefore, is the development of intelligent machines incorporating these oscillators, which involves the problem of control of material to be processed which was touched upon earlier in this report, in addition to the dimensional precision problem.

New materials developed by laser excitation have yet to be commercialized. If a laser generator of considerable output is already developed, based on the research made by the projects, and an appropriate application object is specified, it is assumed that the new materials will be rapidly put into practical use. It is expected that the results of the research introduced in this report, which were obtained by many organizations including the Chemical Engineering Laboratory, will be effectively used in the future.

References

1) H. Irie, S. Tsukamoto, and H. Nakamura; Reports of National Research Institute for Metals, 9 (1988) p.317.

2) I. Masumoto; J. of Jpn. Soc. for Tec. of Plasticity, 27 (1988) p.894.

3) M. Kaneko; Welding Techniques, 33-11 (1985) p.33.

4) S, Kitagawa, H. Nakajima, and S. Sakawa; Preprint of the National Meeting of Jpn. Welding Soc., 35 (1984) p.112.

5) I. Masumoto, T. Shinoda, and H. Ishiyama; Preprint of the National Meeting of Jpn. Welding Soc., 37 (1985) p.276.

6) Y. Arata; J. of High Temp. Soc., 12-1 (1986) p.3.

7) I. Masumoto, and M. Kutsuna; Zis Mitteilungen, 27-3 (1985) p.239.

8) I. Masumoto, M. Kutsuna, H. Iwata, and T. Kawai; Proc. of Int. Conf. on Quality and Reliability in Welding, Sept. 6-8 (1984) A-20-1, Hangzhou, China.

9) I. Masumoto, M. Kutsuna, and M. Hasegawa; Preprint of the National Meeting of Jpn. Welding Soc., 35 (1984) p.296.

10) A. Matsuda; 3rd Plasma Processing Symposium (1986) p.61.

11) P. Sichanugrist, M. Konagai, and K. Takahashi; Extended Abstract 17th Int. Conf. Solid State Devices and Materials, Tokyo, (1985) p.103.

12) S. Nishida, H. Tasaki, M. Konagai, and K. Takahashi; J. Appl. Phys., 58 (1985) p.1423.

13) K. Suzuki, K. Aota, T. Aihara, T. Suziki, K. Kuroiwa, and Y. Tarui; Jpn. J. Appl. Phys., 25 (1986) p.624

14) K. Tachibana, H. Harima, and Y. Urano; Proc. of 47th Jpn. Appl. Phys., (1986) p.876

15) J. Nishizawa; J. Met, 25 (1961) p.149.

16) T. Tamazaki, r. Sugino, T. Ito, and H. Ishikawa; Extended Abstract 18th Int. Conf. Solid State Devices and Materials, Tokyo, (1986) p.213

17) A. Yamada, S. Nishida, M. Konagai, and K. Takahashi; ibid., p.217.

18) S. Nishino, T. Kimoto, K. Furusawa, and H. Matsunami; ibid., p.209.

19) D. J. Ehrlich, R. M. Osgood Jr., and T. F. Deutch; Appl. Phys. Lett., 38 (1981) p.946.

20) J. Y. Tsao and D. J. Ehrlich; Appl. Phys. Lett., 45 (1984) p.617.

21) G. S. Higashi and C. G. Fleming; Appl. Phys. Lett., 48 (1986) p.1051.

22) K. Tsubouchi, K. Masu, and N. Mikoshiba; 33th Joint Symp. of Jpn. Appl. Phys., (1986) p.461.

23) D. J. Ehrlich and J. Y. Tsao; Appl. Phys. Lett., 46 (1985) p.108.

24) V. S. Letokhov, Yu A. Matveetz, V. A. Semchishen, and E. V. Khoroshilova; Appl. Phys., B26 (1981) p.243.

25) A. Ohuchi, H. Moriyama, A. Yabe, et al.; Proc. of Photo Chem. (1987) p.321 (Tohoku Univ.).

26) H. Moriyama, A Ohuchi, A. Yabe, et al.; Proc. of 55th Jpn. Chem. Soc., I (1987) p.171.

27) T. Sugii, R. Ito, and H. Ishikawa; 31th Joint Symp. of Jpn. Appl. Phys., (1984) p.472.

28) T. E. Orlowski and H. Ritchie; Appl. Phys. Lett., 45 (1984) p.241.

29) P. K. Boyer, G. A. Roche, W. H. Richie, and G. J. Collins; ibid., 40 (1982) p.716.

30) M. Murahara, S. Endo, and K. Toyoda; 31th Joint Symp. of Jpn. Appl. Phys. (1984) p.129.

31) M. Tamura and Tokuyama; Oyo Butsuri, 51 (1982) p.152.

32) Y. Horiike, H. Okano, and M. Sekine; ibid., 53 (1984) p.979.

33) T. Ogura, Y. Yamakage, T. Inoue, and M. Hirose; Extended Abstract 18th Int. Conf. Solid State Devices and Materials, Tokyo, (1986) p.205.

34) S. Yokoyama, Y. Yamakage, and M. Hirose; Appl. Phys. Lett., 47 (1985) p.389.

35) A. Katayama; Kagakukeizai; (1985) No.4, p.32.

36) A. Hartford, Jr.; J. Appl. Phys., 51 (1980) p.4471.

37) H. L. Chen and C. Borzileri; IEEE J. Quantum Electron., QE-16 (1980) p.122.

38) Arai, Ishidawa, and Ohyama; J of Laser Soc. of Japan 13 (1985) p.707.

39) M. Murahara, K. Toyoda, and S. Namba; 30th Joint Symp. of Jpn. Soc. Appl. Phys. (1983) p.202.

40) M. Murahara; O plus E., 79 (1986) p.65.

41) Management & Development Intelligence, Diamond, Spring 1988.

Printed in Dunstable, United Kingdom

91230642R00063